Biomass Based Energy Storage Materials

Edited by

Inamuddin[1,2,3], Rajender Boddula[4], Tauseef Ahmad Rangreez[5] and Abdullah M. Asiri[1,2]

[1]Chemistry Department, Faculty of Science, King Abdulaziz University, Jeddah 21589, Saudi Arabia

[2]Centre of Excellence for Advanced Materials Research, King Abdulaziz University, Jeddah 21589, Saudi Arabia

[3]Department of Applied Chemistry, Faculty of Engineering and Technology, Aligarh Muslim University, Aligarh-202 002, India

[4]CAS Key Laboratory of Nanosystem and Hierarchical Fabrication, National Center for Nanoscience and Technology, Beijing 100190, PR China

[5]Department of Chemistry, National Institute of Technology, Srinagar, Jammu and Kashmir-190006, India

Published by **Materials Research Forum LLC**
Millersville, PA 17551, USA

Published as part of the book series
Materials Research Foundations
Volume 78 (2020)
ISSN 2471-8890 (Print)
ISSN 2471-8904 (Online)

Print ISBN 978-1-64490-086-4
eBook ISBN 978-1-64490-087-1

This book contains information obtained from authentic and highly regarded sources. Reasonable efforts have been made to publish reliable data and information, but the author and publisher cannot assume responsibility for the validity of all materials or the consequences of their use. The authors and publishers have attempted to trace the copyright holders of all material reproduced in this publication and apologize to copyright holders if permission to publish in this form has not been obtained. If any copyright material has not been acknowledged please write and let us know so we may rectify this in any future reprints.

Distributed worldwide by

Materials Research Forum LLC
105 Springdale Lane
Millersville, PA 17551
USA
https://www.mrforum.com

Manufactured in the United States of America
10 9 8 7 6 5 4 3 2 1

Table of Contents

Preface

Bone Char as a Support Material to Build a Microbial Biocapacitor
E.D. Isaacs-Páez, B. Cercado.. 1

Nature Inspired Materials for Energy Storage
Nelson Pynadathu Rumjit , Paul Thomas, Shivani Garg, Chin Wei Lai,
Mohd Rafie Bin Johan 21

Biomass Derived Composites for Energy Storage
Pitchaimani Veerakumar and King-Chuen Lin.. 50

Lignin-Derived Materials for Energy Storage
Paul Thomas, Nelson Pynadathu Rumjit, Shivani Garg, Chin Wei Lai,
Mohd Rafie Bin Johan 91

Bamboo Derived Materials for Energy Storage
Sivagaami Sundari Gunasekaran, Thileep Kumar Kumaresan,
Shanmugaraj Andikaddu Masilamanai, Kalaivani Raman,
Raghu Subash Chandra Bose 111

Cellulose-Derived Electrodes for Energy Storage
Shiqi Li, Wenyue Li, Zhaoyang Fan... 124

Keyword Index
About the Editors

Preface

Recently, carbon materials such as graphene, carbon nanotubes, and porous carbons are considered promising energy storage materials due to their physical, chemical, structural, and functional diversity. However, cutting edge science and innovation has impelled the demand for green and sustainable energy storage materials. Biomass is a renewable, eco-friendly, and plenteous resource present on the earth. Most common biomass-derived carbon materials contain nitrogen as well as sulfur, which can be doped by heteroatoms, with the increase in electric conductivity and additional active sites responsible for electrochemical applications. The utilization of low-cost biomass residues as precursor materials to produce carbon-based electrodes for energy storage devices is exploring sustainable and economic pathways. Nature contributes biomass with unusual micro/nanostructures, structural diversities, such as 0D spherical, 1D fibrous, 2D lamellar, and 3D spatial structures.

Biomass is the most renewable and abundant carbon resource, and it shows great potential for sustainable energy production in future technologies. This book provides an in-depth overview of biomass-derived materials for energy storage technologies. The chapters discuss the various types of fundamentals, fabrication methods, energy storage materials from wood, fruits, lignin, seed extracts, and so on. This book also includes bio/nature-inspired materials and activation techniques for battery, water splitting, fuel cells, and supercapacitor technologies. The enhanced sustainable properties of biomass materials are discussed in detail. This book brings together contributions from leading researchers of academia and industry throughout the world. This book is a unique book, extremely well structured and essential resource for undergraduate and postgraduate students, faculty, R&D professionals, production chemists, food chemists, environmental engineers, and industrial experts.

Inamuddin[1,2,3], Rajender Boddula[4], Tauseef Ahmad Rangreez[5] and Abdullah M. Asiri[1,2]

[1]Chemistry Department, Faculty of Science, King Abdulaziz University, Jeddah 21589, Saudi Arabia

[2]Centre of Excellence for Advanced Materials Research, King Abdulaziz University, Jeddah 21589, Saudi Arabia

[3]Department of Applied Chemistry, Faculty of Engineering and Technology, Aligarh Muslim University, Aligarh-202 002, India

[4]CAS Key Laboratory of Nanosystem and Hierarchical Fabrication, National Center for Nanoscience and Technology, Beijing 100190, PR China

[5]Department of Chemistry, National Institute of Technology, Srinagar, Jammu and Kashmir-190006, India

Biomass Based Energy Storage Materials
Materials Research Foundations **78** (2020) 1-20

Materials Research Forum LLC
https://doi.org/10.21741/9781644900871-1

Chapter 1

Bone Char as a Support Material to Build a Microbial Biocapacitor

E.D. Isaacs-Páez[1], B. Cercado[2]*

[1] Instituto Potosino de Investigación Científica y Tecnológica A.C (IPICYT). San Luis Potosí, San Luis Potosí, Mexico

[2]Centro de Investigación y Desarrollo Tecnológico en Electroquímica S.C (CIDETEQ). Pedro Escobedo, Querétaro, Mexico

*bcercado@cideteq.mx

Abstract

Waste biomass can be exploited as an alternative material for carbon-based capacitors. Bone char has intrinsic properties such as surface area, porosity and capacitance, which are suitable for building capacitive electrodes. These properties also favor the development of biofilms to prepare capacitive bioelectrodes. A capacitive bioanode has been constructed with a compost leachate-based biofilm developed on a packed electrode of bone char. The bioanode showed a capacitance of 16.07 $\mu Fs^{(a-1)}/cm^2$ (a = 0.8), a pseudocapacitance of 23.6 $\mu F/cm^2$ and was able to discharge 32,800 C/m^2 in a period of 110 h.

Keywords

Biochar, Bioelectrodes, Biofilm, Capacitance, Charge, Green Energy

Contents

1. Introduction 2

2. Influence of the chemical and textural properties on biochar 5

3. Bioanode preparation 7

4. Accumulated charge 7

5. Biochar-based anode and bioanode capacitances 11

Conclusions 13

Biomass Based Energy Storage Materials   Materials Research Forum LLC
Materials Research Foundations **78** (2020) 1-20   https://doi.org/10.21741/9781644900871-1

Acknowledgements..**14**

List of abbreviations **14**

References...**14**

1.   Introduction

Currently, the replacement of fossil fuels is insufficient to mitigate climate change. Also, the efficient storage of energy for the growing and hungry market of multiple portable electronic devices and hybrid electric vehicles has to be considered. In order to meet the demand for environmentally friendly energy, supercapacitors and electrochemical capacitors have emerged as viable options because they offer higher power density than a conventional battery and higher energy density than a standard capacitor [1]. The energy storage mechanism of a supercapacitor can be divided into two pathways. First, the electric double layer capacitor that is dependent on the electrolyte where ions can access the electrode surface area because the capacitance is only the electrostatic charge accumulated at the electrode-electrolyte interface. The pseudocapacitance is the second form, where the electroactive species are responsible for fast and reversible Faradic processes [2-5].

The physical and chemical properties of an electrochemical capacitor are parameters that strongly affect the energy and power densities. Carbon-based materials are widely used to attain higher energy storage because of their multiple forms (powder, granular, fibers, gels, sheets, composites, monoliths, etc.), controllable porosity, electrocatalytic active sites, ease of processing and relative environmental benignity [4,6-8]. However, tailoring the pore size distribution and the specific surface area (SSA) are crucial for guaranteeing excellent supercapacitor performance. This is appreciable in several works that compared the SSA and specific capacitance (Cm) of various carbon and carbon-based materials. For example, commercial activated carbon has shown high SSA (1000-3500 m^2/g) but Cm values are lower than 200 F/g due to the inaccessible pores, unstable oxygenated groups and inherent resistance of the material. Templated porous carbon with SSA values between 500 and 3000 m^2/g presented Cm values of 120-350 F/g, which indicate good performance. On the other hand, the template technique is limited by its high cost and elaborate procedure. Carbon aerogels are relatively inexpensive to produce and can achieve SSA values similar to activated carbon, but they were reported to show low Cm values compared to carbon nanotubes or graphene [7,9]. Nevertheless, the challenge to obtain a high-performance electrochemical capacitor with these nanostructured materials focuses on the purification issues or the elevated costs of massive production [10]. Li and

Biomass Based Energy Storage Materials   Materials Research Forum LLC
Materials Research Foundations **78** (2020) 1-20   https://doi.org/10.21741/9781644900871-1

co-workers [11] have summarized the commodity prices of carbon materials such as activated carbon ($1250/ton), graphene ($200/kg) and multi-walled carbon nanotubes 50 nm ($700/kg), from which it is clear that nanostructured materials are still expensive for industrial applications.

An alternative carbonaceous material, called biochar, is rising as an economical substitute to activated carbon using low-cost and sustainable precursors, including rice husks, corn straw, bagasse, sludge, animal bones, and garden waste [12-15]; in addition, there is a possibility of feedstock valorization. In contrast to activated carbon, the price for biochar can vary between $0.9/kg and $8.5/kg as a function of the production site [16]. Despite this, biochar remains economically competitive compared to other carbon-based materials.

Biochar has been widely studied to remove different pollutants, such as heavy metals, dyes, pharmaceutical substances, polycyclic aromatic hydrocarbons (PAHs), polychlorinated biphenyls (PCBs), volatile organic compounds (VOCs), and natural organic matter (NOM) [17-26]. However, there have been only a few studies on CO_2 capture and energy storage using this cost-effective and environmentally friendly carbon material [27-29].

This chapter aims to review the recent advances reported in case of biochar for energy storage, either as an electrochemical capacitor or as a bioanode, which is used in bioelectrochemical systems.

Biocapacitors are built via the development of a biofilm on solid supports. This process is initiated by the adhesion of floating bacteria in the medium to a solid surface because of the presence of nutrient-type detritus. Planktonic bacteria are also attached to surfaces by electrostatic interactions [30].

Surfaces with high surface area, porosity and roughness favor bacterial adhesion; therefore, biochar is a suitable support for biofilm formation. Once early colonies of bacteria are established, these spread horizontally and then vertically. The bacteria that form the biofilm produce exopolymers in high amounts such that the bacteria are completely surrounded by a hydrated gel. In the exopolymeric matrix, molecules from the bulk solution are trapped as well as those that are expelled by the bacteria [31].

The biofilm is a dynamic entity; when it is too thick, the inner layers of bacteria may die by starving, thus provoking detachment of the entire biofilm. Erosion by hydraulic flow or particles in movement can also erode the biofilm [31]. In consequence, artificial formation of a biofilm must be carefully performed.

Bacteria clusters grow in the biofilm for protection against environmental factors and for the establishment of ecological beneficial interactions. Among the biofilm activities, the charge transfer between molecules in the cell membrane and the solid supports is of great importance. This natural phenomenon was first described by Lovley for Geobacter species in geological investigations [32]. Recently, even charge transfer has been described for interactions between species [33].

Charge transfer between electroactive bacteria and solid supports follows two recognized mechanisms: direct electron transfer and mediated electron transfer. The former is achieved through direct contact of the cell membrane (and its extensions, such as pili and flagellum) with the solid support [34]. The cell membrane holds catalytic proteins called cytochromes. The types of cytochromes more frequently described for microbial electron transfer are the c-type cytochromes [35,36]; however, the cytochromes may vary as a function of the bacterial genera and species.

The second mechanism for electron transfer in bacteria is based on the transport of charge by redox mediators. Natural mediators are produced by some bacterial species, but external chemical mediators have also been used to enhance microbial charge transfer [37].

Theoretical calculations on the contribution of direct and mediated electron transfer suggest that the prevalent mechanism is direct electron transfer [38]. In this scenario, characteristics of the electrode support material are of key relevance.

The linkage of biochar-based electrodes and electroactive biofilms offers a vast number of possibilities for the construction of bioelectrochemical systems that have not yet been explored. The capacitive properties of biochar-based bioanodes have been attributed to the presence of electroactive biofilms and cytochromes. An increase in the capacitance of electrodes covered by the biofilm compared to uncovered electrodes has been demonstrated for metallic, carbon and polymer-based bioelectrodes [39,40,41].

The capacitive features of biofilms have been investigated in controlled support materials that enable distinguishing the biofilm capacitance from the capacitance of the support material [42]. Genetic modification of bacteria for the overproduction of exopolymeric substances has provided evidence of the role of biofilm thickness on the double layer capacitance [43].

Construction of capacitive bioelectrodes was initially reported for biofilm monitoring studies [44]; later, the capacitive behavior of particular electroactive species, including *Shewanella* and *Geobacter*, was investigated [35,45]. Modifications to the electrode material (e.g. coverage with capacitive layers) have been proposed to increase the electrode capacitance [41,46]. Alternatively, from a sustainability point of view, the use of vegetal biochar as a capacitive electrode to build bioanodes was presented by Katherikeyan et al. [40]. The authors showed that vegetal biomass has cellular structure that resulted in macropores after carbonization; the macropores enabled biofilm growth and promoted the diffusion of nutrients to the biofilm avoiding clogging.

There are multiple advantages of using biochar as an electrode support. First, it contributes to the reduction of pollution by utilizing waste biomass, then allows valorization of wastes, and finally permits a diminution of costs for electrode preparation.

Capacitive bioanodes have been suggested as a way to exploit the energy produced from bioelectrochemical systems. In that direction, microbial fuel cells have been used as devices for the production and storage of energy. Despite the multiple advantages of capacitive bioelectrodes, little research on their preparation and characterization has been done. Therefore, it is necessary to expand research in that direction to explore the enormous combinations of biochar and biofilm sources.

The concept of biochar-based electrodes was investigated by preparing bioanodes from bone char with biofilms from the microbial community in compost leachate. The bone char was characterized; and strategies to prepare the bioanode were developed. The bone char-based bioanode was investigated for capacitance and charge storage. The results confirmed that the capacitive bioanodes were effective alternative for the production and storage of energy.

2.   Influence of the chemical and textural properties on biochar

The textural properties and chemical surface play crucial roles in the performance and efficiency of the materials in their applications either pollutant removal, remediation of soil, CO_2 capture or energy storage.

Biochar is defined as a solid product from the thermochemical conversion of biomass in a zero or low oxygen environment [47]. Commonly, the thermochemical processes include pyrolysis (slow, fast or microwave-assisted), torrefaction, and hydrothermal carbonization [48]. From the literature survey, biochars have been mostly obtained by pyrolysis using temperatures from 300 to 900 °C with a residence time of a few minutes or several hours and a slow heating rate (5 °C/min). Although the modification parameters were similar, the biochar textural properties were different depending on their precursors. For instance, the SSA values of three biochar samples from orange peel, bagasse and cattle bones obtained at 700°C under an inert atmosphere were 501, 388 and 105 m^2/g, respectively [48-52]. In the case of cattle bone char (Fig. 1) the N_2 adsorption isotherm is classified as type IV, which indicated that the bone char was composed of mesopores; also, the hysteresis loop in this isotherm is characteristic of slit-shaped pores

Materials Research Forum LLC
https://doi.org/10.21741/9781644900871-1

[52-54]. The SSA, shape and size pores may be suitable for environmental applications but not for energy storage. Enhancing the capacitance performance requires a high SSA that implies increased microporosity to improve the adsorption capacity between the electrode materials and electrolyte ions, whereas increasing the mesoporosity helps to minimize the ion diffusion distances and facilitate ion transportation [55].

The recent work of Cheng and co-workers [56] has summarized various post modification methods of biochar for electrochemical storage. The most common method to increase the SSA of biochar materials is physical activation with gases such as CO_2, H_2O steam, air or ozone, but the disadvantages are relatively high temperature (> 700 °C) and long activation time. On the other hand, chemical activation, using activating agents such as KOH, $ZnCl_2$, and H_3PO_4, can affect the material mesoporosity and the functional groups on the biochar surface. Their major advantages include augmented SSA (> 2000 m^2/g), a one-step process and the lower energy consumption due to lower pyrolysis temperatures. It is noteworthy that another pathway to modify the biochar chemical surface is by taking advantage of the precursor protein content that can serve as the nitrogen resource to realize heteroatom doping/co-doping. These biochars have been used as supercapacitors and bioanodes in bioelectrochemical systems with notable results for higher power density due to synergistic effects of the surface area, good biocompatibility, and excellent electron transfer [57-58].

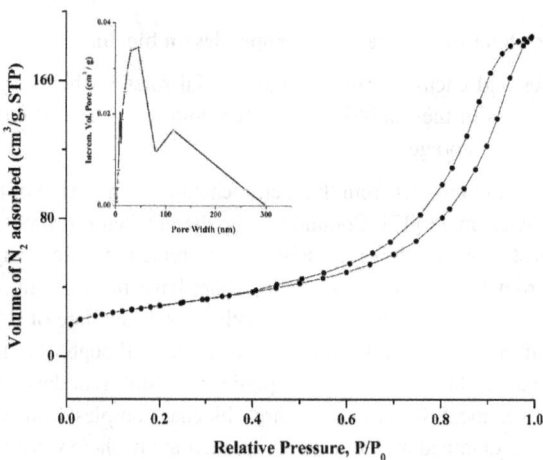

Fig. 1: *N$_2$ Adsorption-desorption isotherm and pore size distributions of biochar from pyrolyzed cattle bone.*

6

3. Bioanode preparation

The bioanodes were formed on biochar over 40 days with successive additions of acetate. One anode was previously immersed in compost leachate for over 30 days. The second anode was submitted to a fixed potential that gradually increased from 0.1 to 0.3 V at day 2, from 0.3 to 0.5 V at day 4, and from 0.5 to 0.6 V at day 10. The open circuit potential was comparatively more stable for the electrode previously immersed in compost leachate, but the maximum voltages for both colonization strategies were very similar reaching to a value of 850 mV within 25 days (Fig. 2).

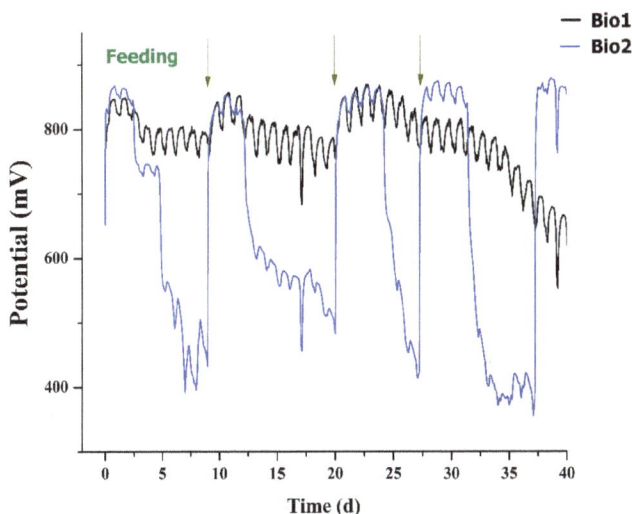

Fig. 2: *Evolution of the open circuit potential of biochar-based bioanode formed in compost leachate. Bio1: bioanode formed at open circuit. Bio2: bioanode formed under electrode polarization.*

4. Accumulated charge

The bioanode was installed in a two-chamber electrochemical cell and submitted to chronoamperometry at a fixed potential of 0.6 V/Ag/AgCl. Successive cycles of production of current showed an increase in the current peaks with the cycles (Fig. 3A); nevertheless, the duration of each cycle differed. In consequence, the released electrical charge also varied.

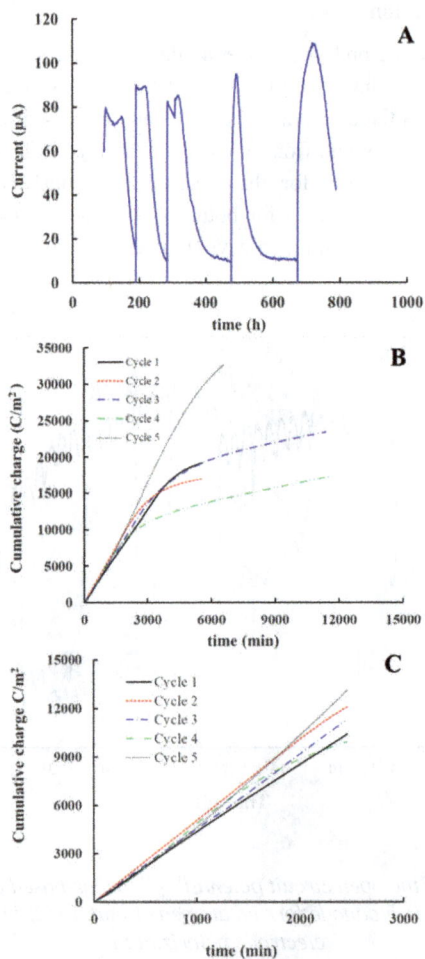

Fig. 3: *Discharging cycles of the bone char-based bioanode at 0.6 V/Ag/AgCl. A) Current curves for successive discharging cycles. B) Cumulative charge for each cycle. C) Cumulative charge before reaching the phase of slowly releasing charge.*

The total electrical charge for each cycle showed that more charge was released in the last cycle (5th cycle) than the previous cycles and that the rate of release of charge declined for the extended cycles (cycles 3 and 4) (Fig. 3B); nevertheless, the cumulative total charge was very similar for the cycles in the early phase, i.e., within 1000 min (Fig. 3C).

Although the bioanode was previously developed on bone char, the charge transfer processes occurred only after the circuit was closed by applying 0.6 V /Ag/AgCl to the anode. Thus an adaptation of the biofilm to electrochemical activity very likely occurred along the successive cycles to give higher amounts of current production and released charge.

The duration of the discharge periods affects the total charge measured; extended periods enable the charge to flow, but with a slower rate since less charge is stored. Electrical current peaks relate to the sum of the instantaneously produced charge and the stored charge, whereas the stable current reveals the capacitive feature for the specifically formed bioanode. Deeke et al. [46] reported a cumulative total charge of approximately 25,000 C/m^2 for a total period of 375 min (15 cycles of 5 min charge and 20 min discharge). The released charge reported by the authors was similar to that obtained in the present work for a longer period (Fig. 3C).

The released charge in the present work was five orders of magnitude higher than the charge measured with a bioanode of *Geobacter sulfurreducens* on graphite bars [59], but it was similar to the charge obtained with a graphite electrode prepared with an additional capacitive layer [46], and was three orders of magnitude lower than that using biochar from carbonized vegetal material [40]. Based on the previous comparisons, it can be concluded that multiple factors greatly affect the charge accumulated in bioanodes; thus, comparisons must be made carefully. Table 1 shows some relevant characteristics for various bioanodes reported in the literature.

Comparatively high released charge has been obtained with modified electrodes of high electrode surface area and with the use of mixed consortium instead of pure cultures. All of these characteristics are a guide to preparing effective capacitive bioanodes. However, not only the materials of the construction of bioanodes affect the amount of charge accumulated and released, but the operational conditions such as charge and discharge rate, and moreover the potential at which the discharge occurs produce effects on the capacitive response of the bioanodes.

The effect of the anode potential at the discharge stage was investigated by Schrott et al. [59]. The authors found that more positive anode potentials resulted in saturation of the maximum released charge due to the redox state of the interfacial cytochromes. The influence of the electrode potential on electroactivity for various microbial species has been reviewed before [60]; nevertheless, for the preparation of capacitive bioanodes from mixed consortia, a screening to determine the optimal bioanode potential might be necessary.

The normalized charge from bioanodes reveals that the exploitation of waste biomass to prepare biochar-based electrodes is clearly promising as an alternative for the construction of biocapacitors. The superior energy storage abilities of the so prepared bioanodes are due to the presence of simultaneous storage mechanisms: charge storage in both the bone char and the biofilm. This condition provided an improved biocapacitor.

Table 1. *Comparative data of the discharge behavior from various bioanodes reported in the literature.*

Solid support electrode	Inoculum source	Electrode surface area (cm^2)	Coulombs at discharging stage (C)	Discharge duration and anode potential[b]	Normalized charge (C/m^2)	Reference
Fluorinated thin oxide (FTO)	Mixed culture from anolyte	Nr	Nr	Nr	5×10^5	[42]
Graphite bars	*Geobacter sulfurreducens*	8	45×10^{-5} to 80×10^{-5}	50 s at 0.4 V	0.562 to 1	[59]
Carbon paper	*Shewanella oneidensis*	3	0.126	30 s at 0.3 V	420	[45]
Polyurethane sponge covered: PANI/rGO/CNT. rGO/CNT. CNT.	Activated anaerobic sludge	4	1036.8 [a]	10, 30, 60 min at -0.1 V	 586-4167 421-2660 82-548	[41]
Plain graphite plate covered with a capacitive layer	Effluent from a microbial fuel cell	22	23.8 to 50.2 [a]	10 to 120 min at - 0.3 V	10,819 to 128,160	[46]
Graphite Biochar from: King mushroom Wild mushroom Corn stem	Mixed consortium from an anaerobic digester	0.25	144 1351 1544 1636	100 h at 0.2 V	$5,760 \times 10^3$ $54,040 \times 10^3$ $61,760 \times 10^3$ $65,440 \times 10^3$	[40]
Bone char	Compost leachate	10.52	34.5	110 h at 0.6 V	32,800	Present work

Nr Not reported
a. Calculated or recalculated with the projected area of electrode taken from data in the materials and methods section in the corresponding reference.
b. Reference electrode Ag/AgCl

With regard to the practical applications, the time for releasing charge from a capacitive bioanode is more extended than using abiotic materials; release of charge in periods on the order of minutes and even hours might accomplish the requirements for the integration of biocapacitors with an electrical grid for buffering current peaks [46].

5.   Biochar-based anode and bioanode capacitances

The electrochemical impedance spectroscopy (EIS) data were fitted to different model circuits. The data from the pristine electrode were fitted with a model including a constant phase element representing the capacitance at the electrolyte-electrode interface. The data from the early and mature bioanode (1-month) were fitted to a model circuit that included the Warburg diffusion element, considering the occurrence of this phenomenon in the biofilm layer [40] (Fig. 4).

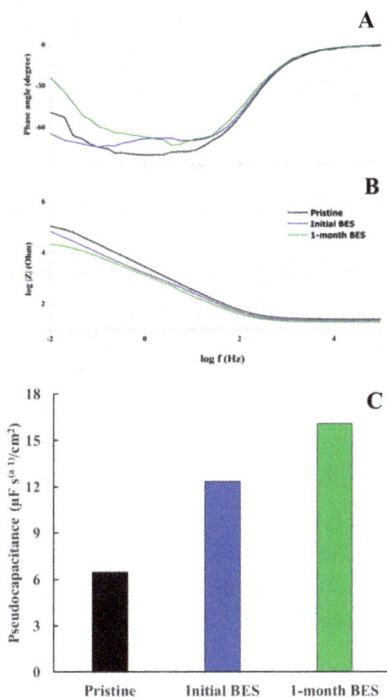

Fig. 4: *Bode plots from EIS data and capacitance for bone char and the bone char-based bioanode at different periods. A) Angle phase. B) Impedance modulus. C) Estimated capacitance.*

The capacitance and pseudocapacitance for the bare electrode were three orders of magnitude lower than that for the bioanode at the startup time, whereas the capacitance for the bioanode with the 1-month biofilm was 23% higher than the capacitance for the bioanode at the start up time, as shown in Table 2.

Table 2. *Model circuits and values of the corresponding elements for the bone char and the bone char-based bioanode.*

	Pristine	Bioanode	Bioanode 1-month
Model circuit	R1+Q2/R2	R1+Q2/(R2+Wd2)	R1+Q2/(R2+Wd2)
Pseudocapacitance (F)	93.65×10^{-6}	0.3882×10^{-3}	$0.248\ 2 \times 10^{-3}$
R1 (Ω)	26.2	22.1	19.4
Q2 (F.s^(a-1))	67.83×10^{-6}	0.1297×10^{-3}	0.1691×10^{-3}
a2	0.89	0.82	0.81
R2 (Ω)	200454	10749	25705
Wd2 (Ω)	---	367982	45907
td2 (s)	---	-14.35	4419

The results in Table 2 confirm that the sum of capacitances of bone char and biofilm resulted in a capacitive bioelectrode that is competitive with the state-of-the-art for bioanodes. Surprisingly, most reported data on biofilm capacitance have not normalized to electrode area, and estimations were made in order to compare the results (Table 3). The calculations showed that the values of biofilm capacitance vary widely; high range reported capacitances as 90 - 666 $\mu F/cm^2$, medium range as 4 - 23 $\mu F/cm^2$, and low range as $0.04 – 1.9\ \mu F/cm^2$.

An extremely high capacitance value was obtained with packed electrodes submitted at electrical current changes. On the other hand, low capacitance was related to the use of pure cultures and small electrode areas.

The biofilm capacitance in the present work was in the medium range and was similar to the value obtained with the use of a mixed culture on FTO [42] and to *G. sulfurreducens* biofilms on gold electrodes [39]. Undoubtedly, the advantages of the bone char-based bioanodes over FTO and gold are the availability in abundance and low cost of the electrode material.

Table 3. *Comparison of the capacitances obtained from bioanodes developed using different support materials and inoculum sources.*

Inoculum and electrode material	Surface area (cm^2)	Capacitance (μF)	Normalized capacitance (μF/cm^2)	Reference
Mixed culture on FTO	Nr	Nr	90 - 5000	[42]
Suspension of an operating microbial fuel cell on carbon cloth	30	10,000 – 20,000 μF for low frequency 45 – 50 μF for high frequency at approximately 1000 μA	333.3-666.7 1.33 – 1.7	[61]
Anaerobic sludge on: graphite rods graphite felt granular activated carbon	Nr	9000 μF 3000μF (inactivated biofilm)	Nr	[62]
Pseudomonas E. coli Geobacter sulfurreducens on gold electrodes	6.45	Over 10 days: 2.5 μF for *Pseudomonas* 12 μF for *E. coli* Over 50 days: *Geobacter sulfurreducens* 500 - 3700 μF	620 0.39 1.9 8.5 - 573.6	[39]
Skimmed milk solution on teflon	8.5	Decreased from 3.7 to 3.2 μF within 2 h	0.04 to 0.38[a]	[44]
Staphylococcus epidermidis on gold chip type sensors	0.016 cm^2	0.22 – 0.68 μF	1.38 – 4.25[a]	[63]
Compost leachate on bone char	10.52	Pseudocapacitance 248.2 Capacitance[b] 169.1	Pseudocapacitance 23.6 Capacitance[c] 16.07	This work

Nr Not reported
a. Calculated from data indicated in the reference.
b. Units μF s$^{(a-1)}$
c. Units μF s$^{(a-1)}$/cm^2

Conclusions

The production and storage of energy are currently in growing need. Technologies for energy storage in capacitors are wide and of great interest; in that sense, biotechnological alternatives are being proposed to substitute high cost and, in some cases, scarce materials for manufacturing capacitors.

Bone char has intrinsic properties such as high porosity and surface area that are suitable for building bioelectrodes. Consequently, bone char has been used as a support material for preparing a bioanode from compost leachate microorganisms. The bioanode has been characterized and presented a capacitance, $(16.07 \mu Fs^{(a-1)}/cm^2)$, pseudocapacitance $(23.6 \mu F/cm^2)$ and release of charge $(32,800 \ c/m^2)$ that are competitive with metallic microbial biocapacitors.

Acknowledgements

This research was financially supported by SEP-CONACYT project 177441 and CONACYT-SENER-FSE project 247006. Dr. Isaacs-Páez benefited by a SENER postdoctoral fellowship.

List of abbreviations

SSA specific surface area
Cm specific capacitance
PAHs polycyclic aromatic hydrocarbons
VOCs volatile organic compounds
NOM natural organic matter
EIS electrochemical impedance spectroscopy
FTO Fluorinated tin oxide
R1 solution resistance
R2 interface resistance
Q2 capacitance
Wd2 Warburg diffusion impedance
a2 non ideality constant

References

[1]   J.P. Zheng, Theoretical energy density for electrochemical capacitors with intercalation electrodes, J. Electrochem. Soc. 152 (2005) A1864-A1869. https://doi.org/10.1149/1.1997152

[2]   B.E. Conway, Electrochemical Supercapacitors, Kluwer Academic, New York, 1999. https://doi.org/10.1007/978-1-4757-3058-6

[3]   P. Simon, Y. Gogotsi, Materials for electrochemical capacitors, Nat. Mater. 7 (2008) 845-854. https://doi.org/10.1038/nmat2297

[4]   A. Ghosh, Y.H. Lee, Carbon-based electrochemical capacitors, ChemSusChem 5 (2012) 480-499. https://doi.org/10.1002/cssc.201100645

[5]   P.J. Hall, M. Mirzaeian, S.I. Fletcher, F.B. Sillars, A.J.R. Rennie, G.O. Shitta-Bey, G. Wilson, A. Cruden, R. Carter, Energy storage in electrochemical capacitors:

designing functional materials to improve performance, Energy Environ. Sci. 3 (2010) 1238-1251. https://doi.org/10.1039/c0ee00004c

[6]   Y. Hernandez, V. Nicolosi, M. Lotya, F. M. Blighe, Z. Y. Sun, S. De, I.T. McGovern, B. Holland, M. Byrne, Y.K. Gunko, J.J. Boland, P. Niraj, G. Duesberg, S. Krishnamurthy, R. Goodhue, J. Hutchison, V. Scardaci, A.C. Ferrari, J.N. Coleman, High-yield production of graphene by liquid-phase exfoliation of graphite, Nat. Nanotechnol. 3 (2008) 563-568. https://doi.org/10.1038/nnano.2008.215

[7]   L.L. Zhang, X.S. Zhao, Carbon-based materials as supercapacitor electrodes, Chem. Soc. Rev. 38 (2009) 2520-2531. https://doi.org/10.1039/b813846j

[8]   A.G. Pandolfo, A.F. Hollenkamp, Carbon properties and their role in supercapacitors, J. Power Sources 157 (2006) 11-27. https://doi.org/10.1016/j.jpowsour.2006.02.065

[9]   Y. Sun, Q. Wu, G. Shi, Graphene based new energy materials, Energy Environ. Sci. 4 (2011) 1113-1132. https://doi.org/10.1039/c0ee00683a

[10]   D.N. Futaba, K. Hata, T. Yamada, T. Hiraoka, Y. Hayamizu, Y. Kakudate, O. Tanaike, H. Hatori, M. Yumura, S. Iijima, Shape-engineerable and highly densely packed single-walled carbon nanotubes and their application as super-capacitor electrodes, Nat. Mater. 5 (2006) 987-994. https://doi.org/10.1038/nmat1782

[11]   S. Li, C. Cheng, A. Thomas, Carbon-based microbial-fuel-cell electrodes: From conductive supports to active catalysts, Adv. Mater. 29 (2017) 1602547. https://doi.org/10.1002/adma.201602547

[12]   Y. Chen, Y.C. Zhu, Z.C. Wang, Y. Li, L.L. Wang, L.L. Ding, X.Y. Gao, Y.J. Ma, Y.P. Guo, Application studies of activated carbon derived from rice husks produced by chemical-thermal process-a review, Adv. Colloid Interface Sci. 163 (2011) 39-52. https://doi.org/10.1016/j.cis.2011.01.006

[13]   T. Taya, S. Ucar, S. Karagöz, Preparation and characterization of activated carbon from waste biomass, J. Hazard. Mater. 165 (2009) 481-485. https://doi.org/10.1016/j.jhazmat.2008.10.011

[14]   S. Nanda, A.K. Dalai, F. Berruti, J.A. Kozinski, Biochar as an exceptional bioresource for energy, agronomy, carbon sequestration, activated carbon and specialty materials, Waste Biomass Valor. 7 (2016) 201-235. https://doi.org/10.1007/s12649-015-9459-z

[15]   D. Angın, T.E. Köse, U. Selengil, Production and characterization of activated carbon prepared from safflower seed cake biochar and its ability to absorb reactive dyestuff, Appl. Surface Sci. 280 (2013) 705-710. https://doi.org/10.1016/j.apsusc.2013.05.046

[16] M.B Ahmed, J.L. Zhou, H.H. Ngo, W. Guo, Insight into biochar properties and its cost analysis, Biomass Bioenergy 84 (2016) 76-86. https://doi.org/10.1016/j.biombioe.2015.11.002

[17] J.H. Park, Y.S. Ok, S.H. Kim, J.S. Cho, J.S. Heo, R.D. Delaune, D.C. Seo, Competitive adsorption of heavy metals onto sesame straw biochar in aqueous solutions, Chemosphere 142 (2016) 77-83. https://doi.org/10.1016/j.chemosphere.2015.05.093

[18] G.N Paranavithana, K. Kawamoto, Y. Inoue, T. Saito, M. Vithanage, C.S. Kalpage, G.B.B. Herath, Adsorption of Cd^{2+} and Pb^{2+} onto coconut shell biochar and biochar-mixed soil. Environ. Earth Sci. 75 (2016) 1-12. https://doi.org/10.1007/s12665-015-5167-z

[19] K.R. Reddy, T., Xie, S. Dastgheibi, Evaluation of biochar as a potential filter media for the removal of mixed pollutants from urban water runoff, J. Environ. Eng. 140 (2014) 04014043. https://doi.org/10.1061/(ASCE)EE.1943-7870.0000872

[20] P. Regmi, J.L Garcia Moscoso, S. Kumar, X.Y. Cao, J.D. Mao, G. Schafran, Removal of copper and cadmium from aqueous solution using switchgrass biochar produced via hydrothermal carbonization process, J. Environ. Manage. 109 (2012) 61-69. https://doi.org/10.1016/j.jenvman.2012.04.047

[21] L. Beesley, E.M. Jiménez, J.L.G. Eyles, Effects of biochar and green waste compost amendments on mobility, bioavailability and toxicity of inorganic and organic contaminants in a multi-element polluted soil, Environ. Poll. 158 (2010) 2282-2287. https://doi.org/10.1016/j.envpol.2010.02.003

[22] Y.C. Li, J.G. Shao, X.H. Wang, Y. Deng, H.P. Yang, H.P. Chen, Characterization of modified biochars derived from bamboo pyrolysis and their utilization for target component (furfural) adsorption, Energy Fuels 28 (2014) 5119-5127. https://doi.org/10.1021/ef500725c

[23] C. Jung, L.K. Boateng, J.R. Flora, J. Oh, M.C. Braswell, A. Son, Y. Yoon, Competitive adsorption of selected non-steroidal anti-inflammatory drugs on activated biochars: Experimental and molecular modeling study, Chem. Eng. J. 264 (2015) 1-9. https://doi.org/10.1016/j.cej.2014.11.076

[24] T. Xu, L. Lou, L. Luo, R. Cao, D. Duan, Y. Chen, Effect of bamboo biochar on pentachlorophenol leachability and bioavailability in agricultural soil, Sci. Total Environ. 414 (2012) 727-731. https://doi.org/10.1016/j.scitotenv.2011.11.005

[25] D. Angın, T.E. Köse, U. Selengil, Production and characterization of activated carbon prepared from safflower seed cake biochar and its ability to absorb reactive

dyestuff, Appl. Surf. Sci. 280 (2013) 705-710.
https://doi.org/10.1016/j.apsusc.2013.05.046

[26]   M.L. Frankel, T.I. Bhuiyan, A. Veksha, M.A. Demeter, D.B. Layzell, R.J. Helleur, J.M. Hill, R.J. Turner, Removal and biodegradation of naphthenic acids by biochar and attached environmental biofilms in the presence of co-contaminating metals. Bioresour. Technol. 216 (2016) 352-361.
https://doi.org/10.1016/j.biortech.2016.05.084

[27]   C. Toro-Molina, R. Rivera-Tinoco, C. Bouallou, Hybrid adaptive random search and genetic method for reaction kinetics modelling: CO_2 absorption systems, J. Clean. Prod. 34 (2012) 110-115. https://doi.org/10.1016/j.jclepro.2011.11.051

[28]   X. Zhang, S.H. Zhang, H.P. Yang, Y. Feng, Y.Q. Chen, X.H. Wang, H.P. Chen, Nitrogen enriched biochar modified by high temperature CO_2–ammonia treatment: Characterization and adsorption of CO_2, Chem. Eng. J. 257 (2014) 20-27.
https://doi.org/10.1016/j.cej.2014.07.024

[29]   A.M. Dehkhoda, E. Gyenge, N. Ellis, A novel method to tailor the porous structure of KOH-activated biochar and its application in capacitive deionization and energy storage, Biomass Bioenergy 87 (2016) 107-121.
https://doi.org/10.1016/j.biombioe.2016.02.023

[30]   M. Hermansson, The DLVO theory in microbial adhesion, Colloids Surf. B 14(1-4) (1999) 105-119. https://doi.org/10.1016/S0927-7765(99)00029-6

[31]   T.R. Garrett, M. Bhakoo, Z.B. Zhang, Bacterial adhesion and biofilms on surfaces, Prog. Nat. Sci. Mater. Int. 18 (2008) 1049-1056.
https://doi.org/10.1016/j.pnsc.2008.04.001

[32]   D.R. Lovley, F.H. Chapelle, Deep subsurface microbial processes, Rev. Geophys. 33 (1995) 365-381. https://doi.org/10.1029/95RG01305

[33]   C.D. Dube, S.R. Guiot, Direct interspecies electron transfer in anaerobic digestion: A review, Biogas Sci. Technol. 151 (2015) 101-115. https://doi.org/10.1007/978-3-319-21993-6_4

[34]   L. Shi, H.L. Dong, G. Reguera, H. Beyenal, A.H. Lu, J. Liu, H.Q. Yu, J.K. Fredrickson, Extracellular electron transfer mechanisms between microorganisms and minerals, Nat. Rev. Microbiol. 14 (2016) 651-662.
https://doi.org/10.1038/nrmicro.2016.93

[35]   P.S. Bonanni, G.D. Schrott, L. Robuschi, J.P. Busalmen, Charge accumulation and electron transfer kinetics in *Geobacter sulfurreducens* biofilms, Energy Environ. Sci. 5 (2012) 6188-6195. https://doi.org/10.1039/c2ee02672d

[36]   A. Esteve-Nunez, J. Sosnik, P. Visconti, D.R. Lovley, Fluorescent properties of c-type cytochromes reveal their potential role as an extracytoplasmic electron sink in *Geobacter sulfurreducens*, Environ. Microbiol. 10 (2008) 497-505. https://doi.org/10.1111/j.1462-2920.2007.01470.x

[37]   F. Kracke, I. Vassilev, J.O. Kromer, Microbial electron transport and energy conservation - the foundation for optimizing bioelectrochemical systems, Frontiers Microbiol. 6 (2015). https://doi.org/10.3389/fmicb.2015.00575

[38]   C.I. Torres, A.K. Marcus, H.S. Lee, P. Parameswaran, R. Krajmalnik-Brown, B.E. Rittmann, A kinetic perspective on extracellular electron transfer by anode-respiring bacteria, Fems Microbiol. Rev. 34 (2010) 3-17. https://doi.org/10.1111/j.1574-6976.2009.00191.x

[39]   N.S. Malvankar, T. Mester, M.T. Tuominen, D.R. Lovley, Supercapacitors based on c-type cytochromes using conductive nanostructured networks of living bacteria, ChemPhysChem 13 (2012) 463-468. https://doi.org/10.1002/cphc.201100865

[40]   R. Karthikeyan, B. Wang, J. Xuan, J.W.C. Wong, P.K.H. Lee, M.K.H. Leung, Interfacial electron transfer and bioelectrocatalysis of carbonized plant material as effective anode of microbial fuel cell, Electrochim. Acta 157 (2015) 314-323. https://doi.org/10.1016/j.electacta.2015.01.029

[41]   H.T. Xu, J.S. Wu, L.J. Qi, Y. Chen, Q. Wen, T.G. Duan, Y.Y. Wang, Preparation and microbial fuel cell application of sponge-structured hierarchical polyaniline-texture bioanode with an integration of electricity generation and energy storage, J. Appl. Electrochem. 48(11) (2018) 1285-1295. https://doi.org/10.1007/s10800-018-1252-9

[42]   A. ter Heijne, D.D. Liu, M. Sulonen, T. Sleutels, F. Fabregat-Santiago, Quantification of bio-anode capacitance in bioelectrochemical systems using electrochemical impedance spectroscopy, J. Power Sources 400 (2018) 533-538. https://doi.org/10.1016/j.jpowsour.2018.08.003

[43]   T. Kim, J. Kang, J.H. Lee, J. Yoon, Influence of attached bacteria and biofilm on double-layer capacitance during biofilm monitoring by electrochemical impedance spectroscopy, Water Res. 45 (2011) 4615-4622. https://doi.org/10.1016/j.watres.2011.06.010

[44]   R. Mauricio, C.J. Dias, F. Santana, Monitoring biofilm thickness using a non-destructive, on-line, electrical capacitance technique, Environ. Monitoring Assess. 119 (2006) 599-607. https://doi.org/10.1007/s10661-005-9045-0

[45]   N. Uria, X.M. Berbel, O. Sanchez, F.X. Munoz, J. Mas, Transient storage of electrical charge in biofilms of *Shewanella oneidensis* MR-1 growing in a microbial

fuel cell, Environ. Sci. Technol. 45 (2011) 10250-10256.
https://doi.org/10.1021/es2025214

[46]   A. Deeke, T. Sleutels, H.V.M. Hamelers, C.J.N. Buisman, Capacitive bioanodes enable renewable energy storage in microbial fuel cells, Environ. Sci. Technol. 46 (2012) 3554-3560. https://doi.org/10.1021/es204126r

[47]   I.B. Initiative, Standardized product definition and product testing guidelines for biochar that is used in Soil. IBI Biochar Standards 2012.

[48]   M.B. Ahmad, A.U. Rajapaksha, J.E. Lim, M. Zhang, N. Bolan, D. Mohan, Y.S. Ok, Biochar as a sorbent for contaminant management in soil and water: A review, Chemosphere 99 (2014) 19-33. https://doi.org/10.1016/j.chemosphere.2013.10.071

[49]   A. Mukherjee, A.R. Zimmerman, W. Harris, Surface chemistry variations among a series of laboratory-produced biochars, Geoderma 163 (2011) 247-255. https://doi.org/10.1016/j.geoderma.2011.04.021

[50]   S.C. Peterson, M.A. Jackson, S. Kim, D.E. Palmquist, Increasing biochar surface area: Optimization of ball milling parameters, Powder Technol. 228 (2012) 115-120. https://doi.org/10.1016/j.powtec.2012.05.005

[51]   A.U. Rajapaksha, S.S. Chen, D.C.W. Tsang, M. Zhang, M. Vithanage, S. Mandal, B. Gao, N.S. Bolan, Y.S. Ok, Engineered/designer biochar for contaminant removal/immobilization from soil and water: potential and implication of biochar modification, Chemosphere 148 (2016) 276-291. https://doi.org/10.1016/j.chemosphere.2016.01.043

[52]   N.A. Medellín-Castillo, R. Leyva-Ramos, R. Ocampo-Perez, R.F. García de la Cruz, A. Aragón-Piña, J.M. Rosales-Martinez, R.M. Guerrero-Coronado, L. Fuentes-Rubio, Adsorption of fluoride from water solution on bone char, Ind. Eng. Chem. Res. 46 (2007) 9205-9212. https://doi.org/10.1021/ie070023n

[53]   C.W. Cheung, J.F. Porter, G. McKay, Removal of Cu(II) and Zn(II) ions by sorption onto bone char using batch agitation, Langmuir 18(2002) 650-656. https://doi.org/10.1021/la010706m

[54]   K.W. Jung, M.J. Hwang, K.H. Ahn, Y.S. Ok, Kinetic study on phosphate removal from aqueous solution by biochar derived from peanut shell as renewable adsorptive media, Int. J. Environ. Sci. Technol. 12 (2015) 3363-3372. https://doi.org/10.1007/s13762-015-0766-5

[55]   Z.S. Wu, Y. Sun, Y.Z. Tan, S. Yang, X. Feng, K. Müllen, Three-dimensional graphene-based macro- and mesoporous frameworks for high-performance electrochemical capacitive energy storage. J. Am. Chem. Soc. 134 (2012) 19532-19535. https://doi.org/10.1021/ja308676h

[56] B.H. Cheng, R.J. Zeng, H. Jiang, Recent developments of post-modification of biochar for electrochemical energy storage, Bioresour. Technol. 246 (2017) 224-233. https://doi.org/10.1016/j.biortech.2017.07.060

[57] T.H. Han, S.Y. Sawant, S.J. Hwang, M.H. Cho, Three-dimensional, highly porous N-doped carbon foam as microorganism propitious, efficient anode for high performance microbial fuel cell, RSC Adv. 6 (2016) 25799-25807. https://doi.org/10.1039/C6RA01842D

[58] X. Han, H. Jiang, Y. Zhou, W. Hong, Y. Zhou, P. Gao, R. Ding, E. Liu, A high performance nitrogen-doped porous activated carbon for supercapacitor derived from pueraria, J. Alloys Compd. 744 (2018) 544-551. https://doi.org/10.1016/j.jallcom.2018.02.078

[59] G.D. Schrott, P.S. Bonanni, L. Robuschi, A. Esteve-Nunez, J.P. Busalmen, Electrochemical insight into the mechanism of electron transport in biofilms of Geobacter sulfurreducens, Electrochim. Acta 56 (2011) 10791-10795. https://doi.org/10.1016/j.electacta.2011.07.001

[60] P.G. Dennis, B. Virdis, I. Vanwonterghem, A. Hassan, P. Hugenholtz, G.W. Tyson, K. Rabaey, Anode potential influences the structure and function of anodic electrode and electrolyte-associated microbiomes, Sci. Rep. 6 (2016) 39114. https://doi.org/10.1038/srep39114

[61] S. Jung, Y.H. Ahn, S.E. Oh, J. Lee, K.T. Cho, Y. Kim, M.W. Kim, J. Shim, M. Kang, Impedance and thermodynamic analysis of bioanode, abiotic anode, and riboflavin-amended anode in microbial fuel cells, Bull. Korean Chem. Soc. 33 (2012) 3349-3354. https://doi.org/10.5012/bkcs.2012.33.10.3349

[62] Z.H. Lu, P. Girguis, P. Liang, H.F. Shi, G.T. Huang, L.K. Cai, L.H. Zhang, Biological capacitance studies of anodes in microbial fuel cells using electrochemical impedance spectroscopy, Bioprocess Biosyst. Eng. 38 (2015) 1325-1333. https://doi.org/10.1007/s00449-015-1373-z

[63] J. Paredes, S. Becerro, S. Arana, Comparison of real time impedance monitoring of bacterial biofilm cultures in different experimental setups mimicking real field environments, Sens. Actuators B 195 (2014) 667-676. https://doi.org/10.1016/j.snb.2014.01.098

Biomass Based Energy Storage Materials
Materials Research Foundations **78** (2020) 21-49

Materials Research Forum LLC
https://doi.org/10.21741/9781644900871-2

Chapter 2

Nature Inspired Materials for Energy Storage

Nelson Pynadathu Rumjit [1], Paul Thomas[1], Shivani Garg[2], Chin Wei Lai [1]*,
Mohd Rafie Bin Johan [1]

1 Nanotechnology & Catalysis Research Centre (NANOCAT), Institute for Advanced Studies
(IAS), University of Malaya (UM), Level 3, Block A, 50603 Kuala Lumpur, Malaysia

[2] Institute of Environmental Studies, Kurukshetra University, Kurukshetra,
Haryana- 136119, India

* cwlai@um.edu.my

Abstract

In our present society, energy depository devices are of great demand. Prevailing energy storing systems are facing challenges in achieving a long-life cycle, higher energy density, biocompatibility and eco-friendliness. Nowadays, nature-derived carbon materials are gaining much research interest in energy repository applications due to their fabrication suitability, economic feasibility and sustainability of many carbons produced from natural precursors which include fruits, plants, microbes and animals. In comparison to human-made carbon nanostructured materials such as carbon nanotubes, graphene and fullerene, nature-derived carbons showed higher capacitance, performance rate and steadiness in supercapacitor applications due to their highly ordered structures and intrinsic nature of nanoporous materials. However, some obstacles persist in the preparation methods to obtain nature-derived carbons with greater carbon yield capacity, energy density and controlled graphite microframeworks. This book chapter is aimed to summarise elemental, chemical compositions and structural-inter relationship charateristics of various nature inspiring materials towards supercapacitor applications. The process for chemical initiation in the enhancement of highly nanostructured nature-derived carbons have been discussed. Additionally, this book chapter discusses future insights for the betterment of nature inspiring carbons for supercapacitor applications.

Keywords

Nature Procured Carbons, Natural Precursors, Consituents, Initiation Methods, Structural-Characteristics Interrelationship, Supercapacitor

Contents

1.   Introduction..22

2.   **Properties of nature-derived carbons properties for fulfilling
the operational need for EDLC- supercapacitors**..25

3.   **Various preparation mechanisms for nature derived carbons
for supercapacitor**...26

4.   **Advantages of naturally-derived carbons over graphene and
CNT for EDLC supercapacitors**...27

5.   **Use of different biological precursors**...28

    5.1   Plant-derived precursors 29

    5.2   Fruit based precursors..31

    5.3   Microbial-based precursors 31

    5.4   Animal-based precursors 32

6.   **Structural characteristics and properties of nature
derived carbons** 33

Conclusions and future directions...38

References 39

1.   Introduction

Energy storing materials such as rechargeable supercapacitors and batteries are regarded as promising candidates for several applications due to their characteristics such as long life cycle, a high-rise in Columbic efficiency as well as energy density and low maintenance cost [1,2]. In our present society, energy storing devices have broad applications in electric vehicles and as portable electronic items. Being in high demand, greater concerns and challenges concerning fabrication cost, eco-friendliness, sustainability, intelligent applications and electrochemical performance of energy devices have arisen [3,4].

Recently, nature-inspired materials are considered as the most effective and innovative techniques in addressing the various challenges faced by energy storing devices. Since, the demands for energy devices are high and availability of mineral resources limited, the overall price of storage devices increased [5]. Existing energy storage devices are non-biodegradable which ultimately result in piling of electronic garbage after their service lifetime.

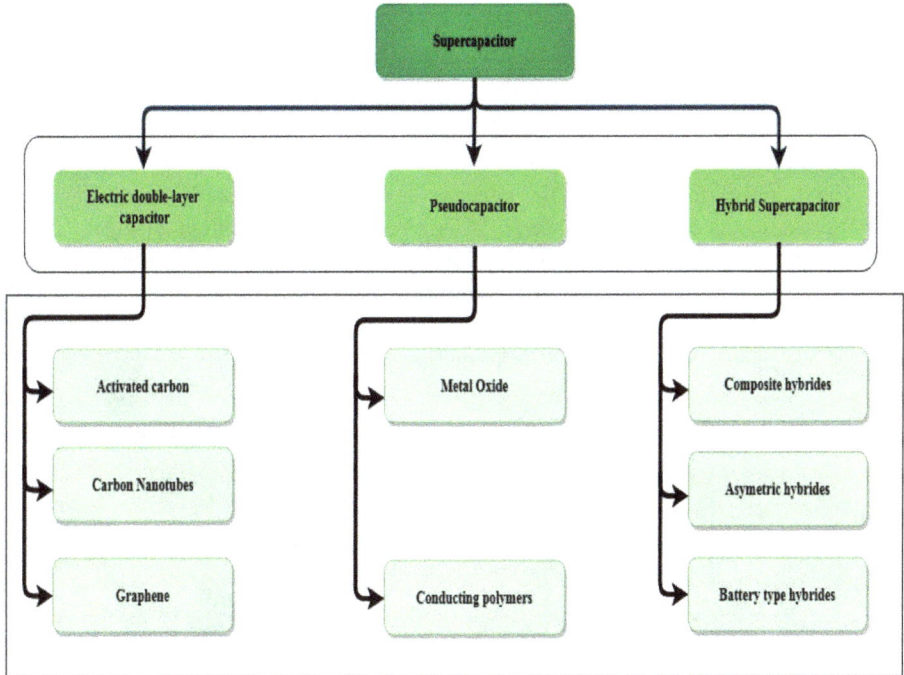

Figure 1. Classification of supercapacitors based on energy storage[12].

Hence, it is highly preferred to identify low-cost, renewable and eco-friendly materials for substituting existing components in storage devices [6,7]. A billion years ago, storage systems and energy conversion have developed from nature and are found to be well adapted to metabolism, growth and distribution of energy in various life forms [8]. Recently, natural materials such as biobased electrodes have been fabricated from renewable biomass sources [9].

The mechanism involved in the functioning of electron shuttles in extracellular electron transport is through the changeable redox-cycling process [10]. Recently, supercapacitors utilising redox biomolecule as functional Faradaic materials exhibited superior energy density compared to transition metal derived supercapacitors [11].

Based upon the energy depository process, supercapacitors are categorised into three types as described in **Figure 1**. The first one is the electrochemical double layer capacitor commonly known as (EDLC). Examples of electrolytes used between the electrodes are potassium hydroxide and sulphuric acid. During the mechanism, Helmholtz double layer formation occurs between conductive electrolyte/electrode interfaces which results in rapid sorption/desorption of electrolyte ions favouring a greater rise in power density by the non-Faradaic process.

The overall performance of EDLC depends on the thickness of the Helmholtz layer and the size of the electrode surface [12]. As a result, carbon materials of highly porous nature and surface characteristics such as carbon nanotube (CNTS), activated carbon (AC) (500-3000 m^2/g)[13,14] and graphene have been extensively used as electrode materials for industrial applications. Figure 2 illustrates a basic schematic diagram of electrochemical double layer supercapacitor.

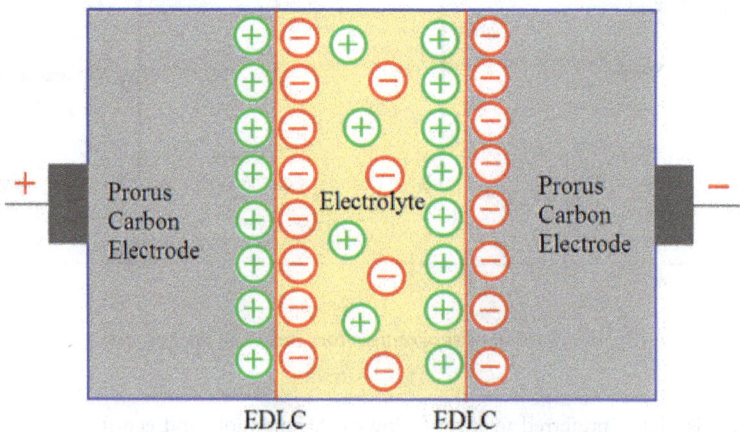

Figure 2. *Diagramatic representation of electrochemical double layer supercapacitor [12].*

Pseudocapacitors come under the second type of supercapacitors. These are made utilising metallic sulfides (S^{2-})/hydroxides (OH^-)/oxides (O^{2-}) or with N and O functional group compounds. They function by using rapid, revocable Faradaic redox reactions occurring at the surface of the electrode.

Resembling EDLC, chemical reactions take place at the electrode surface generating a high energy density at a low capability rate, less operating voltage and short life cycle [15]. The hybrids of high power density EDLC electrode and high energy pseudocapacitive electrode which come under the third category of supercapacitors exhibited excellent energy storage material applications with better cycling stability. Li-ion hybrid capacitors are the recent examples of hybrid type supercapacitors [16].

Another drawback faced by the present energy repository systems is the laboratory development formulations of storage devices which require intense experimental conditions. In a natural system, with the help of biological templates, the assembly of complex micro-organelles is explicitly controlled. By utilising suitable bio templates, easy preparation of energy storage materials has been achieved under mild experimental conditions [17]. Nature-derived carbon materials have been produced by the conversion of biological products such as plants, animal waste and foodstuff microorganisms into the highly porous structure using activation and thermal carbonisation processes [18].

During the carbonisation process, the biomaterials undergo heating under high temperature and inert gas conditions resulting in the formation of highly poriferous carbon structure [19]. Upon subsequential activation, the carbon structure is configured to highly interconnected 3-D structures exhibiting properties such as high conductivity, porosity and surface area, making them a superb candidate for supercapacitor storage applications [20].

Compared to graphene and CNTs, nature-derived carbon materials store charges by use of electrical double layer (EDL) mechanism [21]. Including nature derived carbons, supercapacitors which utilise EDL process for charge depository are categorised as electrochemical double layer supercapacitors [22]. This chapter summarises the recent advances in utilising natural materials for supercapacitors which are of high demand in energy depository applications.

2.   Properties of nature-derived carbons properties for fulfilling the operational need for EDLC- supercapacitors

- The existence of high surface area and significant distribution of pore size signifies greater performance rate and EDL capacitance.

Biomass Based Energy Storage Materials   Materials Research Forum LLC
Materials Research Foundations **78** (2020) 21-49   https://doi.org/10.21741/9781644900871-2

- High specific capacitance conforms to high energy density and power.

- Lesser equivalent series resistance (ESR) results in more availability of power with lower power loss and voltage drop.

- Nature-derived carbons with long lifetime and cyclable condition signify a high rise in the stability of electrode under repeated charging-discharging cycles and altered temperature conditions.

At present nature-derived carbons are gaining vital importance due to their fantastic applications in the absorption of water contaminants [23], ionic liquids [24], CO_2 capture [25], lithium-sulfur battery [26] and significant utilisation in supercapacitors [27].

Nanosheets and poriferous carbon flakes obtained from orange peels [20], dry elm samara [24] and human hair [23] exhibited high specific capacitance more than 400 Fg^{-1} when compared to other EDLC-type supercapacitors such as CNT [28] and carbon nanofibers (CNF) [27]. Nature-derived carbon materials from gelatin [29], chicken egg-shell membrane [30], chitin [31] and auricularia [32] showed high conductivity of about 1.12×10^3 Sm^{-1} with less ESR.

Porous nanocarbon sheets formed from poplar and willow catkins [33,34] were doped with (N,O,B,S) resulted in high performance rate showing 68% retention capacity at a higher current density about 30-50 A g^{-1}. Activated carbon(AC) procured from gelatin [29] and chicken egg white [35] showed high retention capacitance of about 113% and 79.2%, and cycling stability exceeded over 15,000 cycles. While nano carbon sheets obtained from dry elm samara [24] showed a retention capacitance of about 98% with cycling stability over 50,000 cycles.

3. Various preparation mechanisms for nature derived carbons for supercapacitor

To synthesise nature derived carbons for supercapacitor applications with organised pore size distribution (PSD) and better yield capacity in an economic and simple manner, immense efforts have been carried out in enhancing the overall processes of supercapacitor fabrication from the selection of precursors, carbonisation, activation and electrochemical characterisation.

Commonly used precursors for nature-derived carbons are wheat, rice straw [36,37] poplar catkins [38] and pistachio [39] due to their bulk availability and less cost. Chicken shell membrane, kombucha [40] and orange peels were chosen due to their unique ordered porous structures while *B. subtilis* [41] and chitin [31] were selected due to the abundant availability of heteroatoms (N, S and O).

In order to enhance functioning characteristics of nature-derived carbons, active materials such as boric acid (H_3BO_3) [29], graphene oxide (GO) [41] and melamine ($C_3H_6N_6$) [37], were blended with biological precursors before the activation/carbonisation mechanisms in order to perform as a good conducting microstructured constituents, or as a source for heteroatom and for creating rigid templates for 2D and 3D structure long-range order. The routeways of activation and carbonisation are significantly depend on physical and chemical nature of biomass precursors. For example, from rice straw, black liquor has been synthesised by the KOH solvent to derive the biomass material by hydrothermal process [36].

The black liquid on drying acquired black composite which further undergo activation at 800°C and carbonisation at 400°C under N_2 atmosphere. While *B. subtilis* bio-based precursor was blended with various activating agents such as KOH and $ZnCl_2$ and undergo carbonisation at 800°C for two hours under N_2 atmosphere for fine-tuning of N and O elements in the precursor [42].

Initially in the case of human hair, a fine cutting step was carried out about 5 mm long and further precarbonised at 300°C prior to the activation process [23]. Finally, the electrochemical capacity performance of biomass carbon was evaluated by characterising in aqueous electrolytes. Examples of commonly used electrolytes include 6M KOH, $LiPF_6$ (Lithium hexafluorophosphate) and $C_6H_{11}BF_4N_2$ (1-ethyl-3-methylimidazolium tetrafluoroborate) [43,44].

4. Advantages of naturally-derived carbons over graphene and CNT for EDLC supercapacitors

- The precursors obtained from nature-derived carbons are cost-effective and abundantly available, which are sourced from plants, microbes, animals and food wastes.

- The overall structure of biological macromolecules is well preserved during the conversion to carbon under the protection of inert gas, resulting in the formation of highly interlinked porous nanocarbon structures with high conductivity.

- Different types of biological precursors can be transformed into nature-derived carbons through various steps such as carbonisation, activation and purification. Furthermore, introduction of metallic compounds during the conversion process further enrich nature-derived carbons with remarkable catalytic property and electrochemical capacity.

Biomass Based Energy Storage Materials   Materials Research Forum LLC
Materials Research Foundations **78** (2020) 21-49   https://doi.org/10.21741/9781644900871-2

- The synthesis of nature-derived carbons does not require any extreme experimental conditions when compared to graphene and CNTs. Subsequently, nature-derived carbons are eco-friendly and energy saving with high-performance rates therefore they are highly applicable for energy storage.

5.   Use of different biological precursors

Biomass sources for precursor selection are commonly categorised into four types, i.e., plant, animal, food and microbial-based biomass as shown in Figure 3. To produce nature-derived carbons with high yield capacity, porosity and conductivity for supercapacitor applications some criteria have to be followed as listed below:

- The biomass for precursors selection should be highly interlinked and stable under high-temperature conditions. Examples include chitin, lignin and keratin used for improved carbon yield and improved rate formation of aromatic carbon during the carbonisation process.

- The biomass precursors containing non-crosslinking, aliphatic and low MW compounds contribute to carbon yield of minor importance and hinder aromatic carbon formation by producing volatile compounds that restrict the flow and fusion.

- The biomass precursors containing higher O_2 constituents hinder the formation of aromatic carbon and further escalate biochemical oxygen demand (BOD)[45]; while the presence of nitrogen constituents enhance the conductivity of precursor[46].

Biochar obtained from chicken egg shell membrane exhibited high super capacity features such as high specific capacitance, conductivity and long cycling stability [30]. However, the yield is significantly less (0.2–0.25 g/per chicken egg) [48], irrespective of loss in weight throughout the activation and carbonisation, which make the overall production procedures less economic and more labor consuming.

A similar situation was observed for biochar obtained from willow catkins. Even though this precursor showed amazing super-capacity properties [34], the final yield of carbon was found to be lower after the activation of about 5.5 wt %, which was significantly less when compared to other natural biomasses such as okara [49] (52.1 wt %), lignin [50] (40.3 wt %), and rice straw [36] about (24.8 wt %).

Materials Research Forum LLC

https://doi.org/10.21741/9781644900871-2

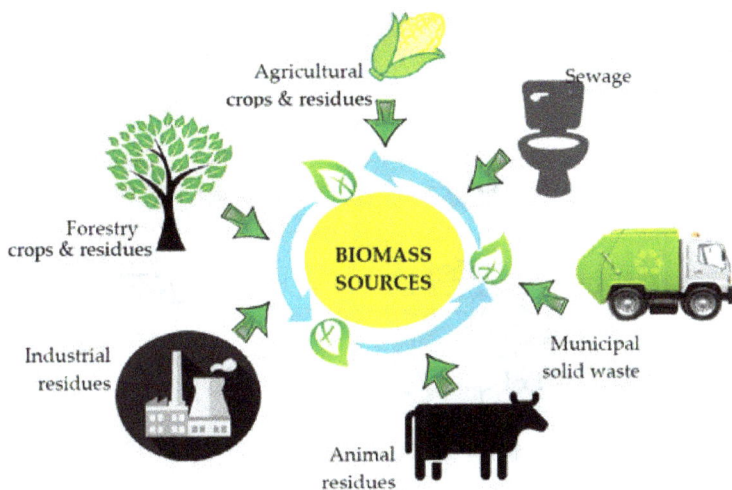

Figure 3.Various sources of biomass raw materials [47].

5.1 Plant-derived precursors

Plant-derived biomass has high carbon fractions ranging from 40-60%. During the conversion process, the actual carbon yield depends on various constituents, particularly the mass proportions of cellulose, polyose and lignin as shown in Table 1. Lignin exhibits the highest thermal stability and shows a pivotal role in the formation of activated carbon and chars [51].

For cellulosic and polyose, the carbon yield was lower compared to lignin which resulted in low carbon content in final chars. For cellulose, about 78% weight loss was observed at 400°C for 1h [52].

Plant precursors having higher cellulose content enhanced BOD of biochars which resulted in the formation of non-aromatised carbon [53]. There was no considerable interaction between cellulosic and ployose during pyrolysis at 500°C, and no noticeable interaction was found between lignin and cellulose constituents of wood biomass [54].

Hence, it is evident that plant-derived biomass should be selected based on characteristics such as low cellulose-high lignin fractions, high N and low O fractions to produce plant derived biomass with high carbon yield, better conductivity and graphitisation for supercapacitor applications. Recently, cellulose procured carbon nanospheres (CCS) were obtained from corn straw by hydrothermal carbonisation and has been found to be applicable as anode components for Li-ion batteries.

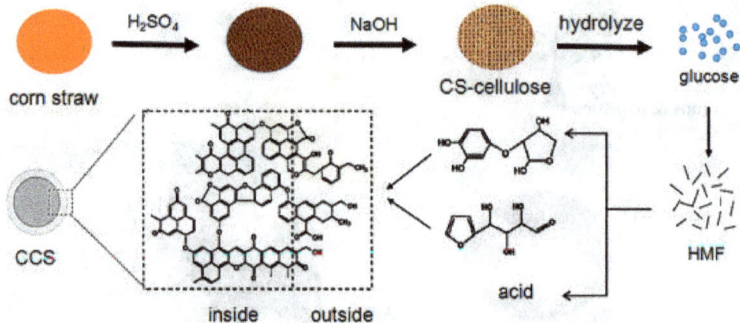

Figure 4. *Schematic illustration of the process involved in CCS preparation [55].*

After 36 hours of thermal carbonisation, an excellent specific capacity of 577 mA h/g has been achieved and showed retaining cycling stability of 100 cycles at 0.2C. CCS enhanced the surface area and boosted the specific capacity of lithium-ion batteries by providing more effective sites for the interpolation of lithium ions. Hence green synthesis of LIBs succeeded.

Table 1. *Weight(%) of cellulose, hemicellulose and lignin to carbon yield for various plant biomass [51].*

Plant waste	Cellulose (%)	Hemicellulose (%)	Lignin (%)	Yield after conversion (%)	Yield after conversion (%)
Coconut shell	8.8	24.5	66.7	30.7	25.5
Olive stone	10.5	14.3	75.2	25.2	31
Plum pulp	5.5	15.5	78.9	22.1	25.8
Plum stone	14.1	15.0	70.9	31.2	24.7
Coconut shell (synthetic)	8.6	24.4	66.8	33.6	32.5

5.2   Fruit based precursors

Food derived biomass contains qualitative components such as lipid (1.4-28.6 %), ash, crude protein (2.8-18.06%), crude fibers and carbohydrates [56,57]. The higher proportions of lipid and crude protein present in fruit biomass limit the overall carbon yield during the thermal carbonisation process due to the degradation of crude protein and lignin at the temperature less than 300°C and finally release VOCs such as H_2O, NH_3, CO_2, methyl ester and olefines[20].

However, lipid and crude protein contain high N and Phosphorus which support the generation of heteroatom doping carbons. The crude fibres of fruit derived precursors are commonly more abundant in cellulosic components than lignin which results in the inadequate generation of biobased carbons with better yield and structural characteristics [58].

5.3   Microbial-based precursors

Nature-derived carbons are obtained from microorganisms such as bacterial cellulose and fungus due to their abundant nature availability and higher rate of growth. Major constituents for microbial-based precursors are carbohydrates, crude (fiber, fat, protein) and ash [59]. In microbial based biomass, the carbohydrates are composed of chitins which are crosslinked with glucan and provide a primary carbon source during the carbonisation process [60].

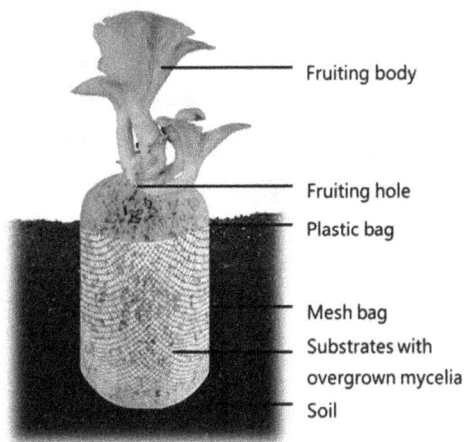

Figure 5. Fruiting body and mycelia formation in mushrooms [61].

Mushrooms are one of the potential precursors used to synthesise natured-derived carbons in the list of microbial-based biomass. The sporocarp (fruit body) of the mushroom is commonly used for the production of biobased carbon than mycelia due to the higher presence of nitrogen content in the of range 3-17% [62] as shown in Figure 5. Hence mushrooms are optimistic precursors to generate N-doped biobased carbons and applicable for energy storage.

5.4   Animal-based precursors

Commonly used animal-based precursors for the production of biobased precursors are molluscans, insects and crustaceans [63]. Chitin is the chief component present in these biomasses and found suitable for generation of biobased carbons due to high content of nitrogen(6%) [64], high yield of carbon (4-40%) [65,66] better thermal and chemical stabilities compared to cellulose, and high interlinking between chitin-catecholamine and chitin-glucan complex [67,68]. Chitin is abundantly found in animal biomasses such as crab (14-72%), lobster (60-75%) and butterfly (64%) [63]. Apart from this claws, hooves, hair and horns of animals contain highly structural fibrous proteins named as keratin which exhibit similar properties compared to chitin [69,70].

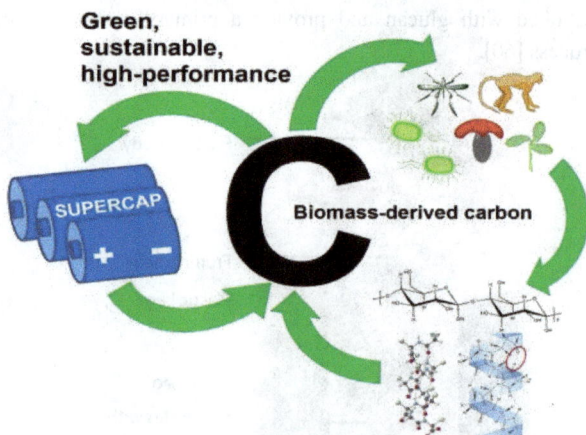

Figure 6. *Nature-derived carbon precursors for supercapacitor applications [20].*

Keratin is categorised into 2 types (α and β keratin) composed of amino acids. As a result of strong covalent bonding (disulfide-cystine interlinking) and non-covalent (H bonding) occurring within the structure of the molecule, keratin displays adequate stability during the carbonisation process and producing biobased carbons with high yield capacity.

One of the remarkable examples is the biobased precursor derived from human hair which exhibited greater surface area, cycling stability for long-term, as electrode components for supercapacitors.

6.   Structural characteristics and properties of nature derived carbons

The features and structural characteristics of biobased carbons are enhanced by employing activation agents such as KOH and ZnCl$_2$ [71]. During physical activation, gasifying products evolved such as CO_2 and H_2O enhanced the formation of highly porous carbon structure which employs interaction between C and K ions. Interpolation of metallic K into carbon framework expanded the graphitic carbon network phase, and exfoliation of nanocarbon structure takes place [72]. After KOH activation, the surface area characteristics of nature procured carbon raised extremely (10-100 fold) by the formation of nanosheets with high microporosity [73].

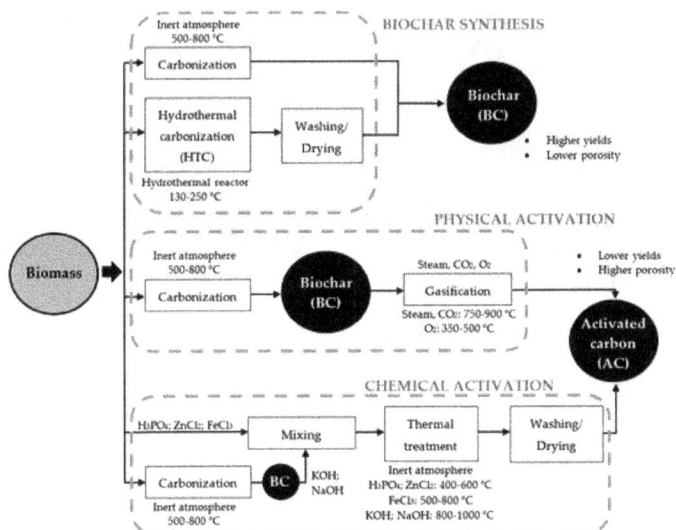

Figure 7. Various synthesis routeway for Activated carbon and biochar from biomass resources [47].

For supercapacitor applications, nature-derived carbons acquired from low KOH/C ratio (1-3) and activation temperature range (700-900°C) are most preferred due to their higher surface area, better performance rate and specific capacity [74]. The activation process causes shrinkage of carbon biomass mesopores (5nm) and integrates to form highly interlinked meso or macroporiferous structure [71]. At higher KOH/C mass ratio, expansion of micropores takes place in carbon architecture and integration of micropores is hindered instead of forming highly interlinked meso or macroporiferous structures. KOH initiation combined with physical activation agents such as CO_2 also with melamine and thiourea generated biochar with organized functional groups, better doping and enhanced supercapacitor material properties [34,75].

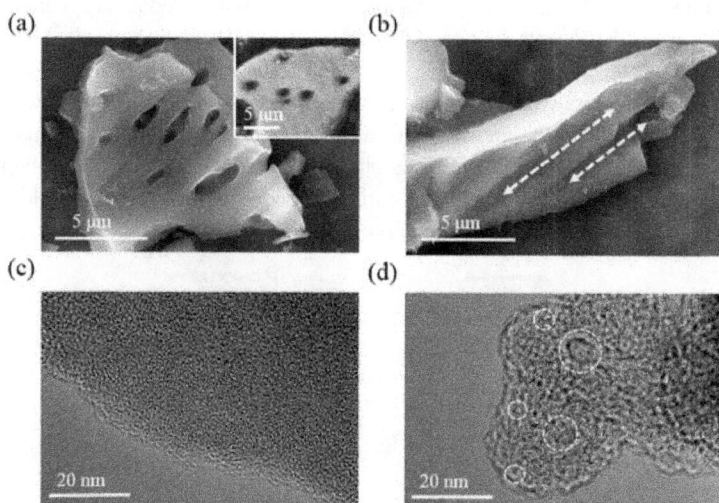

Figure 8. SEM (a and b) and TEM (c and d) pictures of pyrolysed activated carbon derived from corncob [76].

Recently carbon constituents procured from corncob biomass utilising 6M KOH as an activating agent have been found suitable for supercapacitor application as electrode components. A specific capacitance of 299F/g was obtained at power density (500W/kg) with retention capacity of 99.9% achieved even after 4000cycles. SEM and TEM results showed a greater BET surface area of 1471.4 m^2/g as shown in Figure 8 signifying a highly porous nano carbon framework.

Apart from KOH, $CaCl_2$ and $ZnCl_2$ were also used as activation agents for synthesis of nature-derived carbons. However, $ZnCl_2$ was not very productive in synthesising microporous carbon due to the larger Zn^{2+} radius. The combined effect of Zn^{2+} and C created the formation of amorphous carbon frameworks which resulted in the decline of the material performance rate of supercapacitors [77]. Hence, KOH as activation agent has been commonly preferred for supercapacitor components. Recently, KOH activation along with chemical blowing combination method was used to synthesise ultraporiferous carbon nitride(N^{3-}) structures which exhibited excellent super capacitive characteristics[78].

Table 2. *Various nature derived precursors obtained from biomass, structural specifications and application towards the performance of supercapacitors.*

Various precursors	Productio n of nature derived from carbon.	Initiation method	Surface area (m^2/g)	Specific capacitance (Fg^{-1})	Current density (Ag^{-1})	Cycling stability	Performance rate	Ref.
Coconut fibers	Activated carbon(A C)	KOH:C= 1:4, 700^0C for 1h	2898	266.2	0.1	84% after 10000 cycles	76.2% at 100(A/g)	[79]
Corncob residues	Highly porous carbon	6MKOH, Steam activation, 850^0C for 45min	1210	314	1	82% after 100000 cycles	Energy density 15Wh/kg, Yield capacity- 23.2%	[80]
Wheat flour	Highly ordered N-doped carbon	KOH:C= 1:1, 800^0C for1h	1294	383.4	1	91.6% even after 5000 cycles	65.1% at 10 (A/ g)	[81]
Char from bamboo	Poriferous graphitic biobased carbon	K_2FeO_4 (Potassium ferrate):C= 2:1, 800^0C for 2 h	1732	222.6	0.5	84% even after 5000 cycles	51.7% at 20(A/g)	[82]
Crab shell	Highly ordered poriferous carbon	KOH:C= 1:3, 800^0C for 1h	1905	322.5	1	94.5% even after 10000 cycles	65.5 % at 10(A/g)	[83]

Cotonier strobile	Poriferous AC tubes	6MKOH, 800^0C for 200-300s	2670	346	1	Nearly 100% even after 10000 cycles	84.2% retention capacity	[84]
Camelia olifera shell	AC	One step carbonisation,$ZnCl_2$:shell =3:1,630^0C 1h	2851	146	0.5	86% even after 3000 cycles	86.5% at 4 (A/g)	[85]
Fig fruit inside portion	Highly poriferous foam-like carbon framework	One step carbonisation, KOH:C= 3:1, 900^0C for 2h	2337	340	0.5	Greater than 100% even after 10000 cycles	50.1% at 50(A/g)	[86]
Banana skin	N-doped poriferous carbon foam	Activation not done	1357	210.7	0.5	100% even after 5000 cycles	79.8% at 50(mV/s)	[87]
Orange skin	3D nano poriferous carbon	KOH:C= 3:1,600^0C for 1h	2160	460	1	98% even after 10,000 cycles	59.2% at 100(A/g)	[88]
Kombucha	Highly ordered poriferous carbon	KOH/C= 6:1,700^0C for1h	917	326.2	1	91.3% even after 500cycles	82.1% at 20(A/g)	[40]
Bradyrhizobium japonicum(Bj)	N-doped 3D ordered poriferous carbon	One step carbonisation,$ZnCl_2$:C =1:5 800^0C for 3h	1275	358	1	91% even after 8000 cycles	83.7% at 20(A/g)	[89]
Garlic skin	3D highly ordered poriferous carbon	One step carbonisation,KOH:C =4:1 800^0C for 2.5h	2818	427	0.5	94% even after 5000 cycles	315 F/g maintained at 50(A/g)	[90]
Indian cake rusk	Highly ordered nano-poriferous AC	KOH:C= 1:3,750^0C for 3h	1413	381	1.7	95% even after 6000 cycles	Energy density 47.2(Wh/kg) also, power density 22645(Wh/kg)	[91]

Pea skin	N-doped AC	KOH:C= 1:1,700^{0}C for 2h	1828	141	1.3	75% even after 5000 cycles	Specific energy 19.7 (Wh/kg) and specific power 25.5 (kW/kg)	[92]
Sugar industry waste	Highly poriferous carbons synthesised.	Calcination process at 800^{0}C for 4h ZnCl$_2$:C= 1:1	900	120	0.2	About 100% even after 1000 cycles.	The power density of 4.13 (Kw/kg)	[93]
Sawdust waste	Highly porous AC fibers	KOH:C= 1:6, 850^{0}C for 2h	1013	225	0.5	94.2% even after 10000 cycles	85.2% at 10(A/g)	[94]
Soybean root	3D highly ordered poriferous carbon	KOH:C= 3:4,900^{0}C for 2h	2143	276	0.5	98% even after 10000 cycles	Energy density 100.6 (Wh/kg)	[95]
Rice Husk	AC with high surface area characteristics	KOH:C= 5:1,850^{0}C for 1h	2696	147	1	90% even after 10000 cycles	The energy density of 5.12 (Wh/kg)	[96]
Hybrid of nickel cobaltite nanograss and lemon skin waste	Highly poriferous carbon synthesised.	One step carbonisation 5g of carbonised sample initiated with 20% of KOH, 800^{0}C for 3h	1012	17.5	1	92% retention after 3000 cycles	Energy density 6.60 (Wh/kg)	[97]
Eucalyptus leaves	N-doped poriferous nano carbon sheets	KHCO$_3$:C = 1:4 850^{0}C for 5h	2133	71	2	97.7% retention after 15000 cycles	Specific capacity 818 (mAh/g)	[98]
Human hair	Heteroatom (N, O, S) doped poriferous carbon flakes	KOH/C = 2:1, at 800^{0}C for 2 h	1306	445	0.5	98% retention after 20000 cycles	51.2% at 10(A/g)	[23]

Conclusions and future directions

To summarise the chapter, nature-derived carbons exhibited superb performance rate characteristics towards supercapacitor applications which are more green, economic and eco-friendly in nature. The overall performance of nature-procured carbon is highly dependent on the structural features at the subatomic and elementary levels. Hence it necessitates a coherent design in various procedures such as choice of precursors, activation and carbonisation to achieve remarkable rate performance for supercapacitors with optimised surface features, conductivity and specific capacity.

- Even though nature derived carbons show amazing supercapacitor performances when compared to conventional precursors such as CNTS and CNF (carbon nanofibrils), various hurdles are faced by nature derived carbons.

- The elemental and chemical constituents of various biobased precursors vary between 5.5-52.1 wt%. However, the low carbon yields for most of the biomasses are not favorable for superb capacitor applications and thus limit the suitability of precursors for wide-scale production. This can be overcome by enhancement of biomass precursors during carbonisation and activation using activating agents KOH, organic combinations and metallic facilitators (eg.$ZnCl_2$).

- Nature-derived carbons showed limitations in EDL capacitors due to low energy density and specific capacity. This can be overcome by the introduction of heteroatom (N, B, O and S) doping into nature derived carbon prior to or after carbonisation. Energy densities of nature acquired carbons have been furthermore enhanced by coating with metal oxides(MO) and conducting biopolymers.

- At present the production and generation of nature-derived carbons show inadequate engineering controls over graphitisation and carbonisation methods. This results in irregular arrangements of macro/meso pores over graphitised and carbonised structures. This can be overcome by the use of a blowing agent such as NH_4Cl. Addition of blowing agents during the carbonisation process attained graphene walls with highly ultra-microporiferous carbon frameworks.

Also, structural nanotemplates, such as silica, resins and organic-metal structures, can be blended with nature-derived precursors during the process of carbonisation to achieve highly poriferous carbon form with well-arranged dispensation of pore size. Apart from activation by KOH, other initiation methods such as hydrothermal and air initiation techniques can be used to acquire nature-derived carbons with enhanced N constituents contents with better self-supporting characteristics.

Biomass Based Energy Storage Materials	Materials Research Forum LLC
Materials Research Foundations **78** (2020) 21-49	https://doi.org/10.21741/9781644900871-2

Acknowledgements

This research work was financially supported by the University Malaya Impact-Oriented Interdisciplinary Research Grant (No.IIRG018A-2019) and Global Collaborative Programme - SATU Joint Research Scheme (No. ST012-2019).

References

[1]	N. Nitta, F. Wu, J.T. Lee, G. Yushin, Li-ion battery materials: Present and future, Mater. Today 18 (2015) 252–264. https://doi.org/10.1016/j.mattod.2014.10.040

[2]	L. Peng, Y. Zhu, D. Chen, R.S. Ruoff, G. Yu, Two-dimensional materials for beyond-lithium-ion batteries, Adv. Energy Mater. 6 (2016) 1600025. https://doi.org/10.1002/aenm.201600025

[3]	Y. Zhang, Y. Zhao, J. Ren, W. Weng, H. Peng, Advances in wearable fiber-shaped lithium-ion batteries, Adv. Mater. 28 (2016) 4524–4531. https://doi.org/10.1002/adma.201503891

[4]	W. Weng, Q. Sun, Y. Zhang, S. He, Q. Wu, J. Deng, X. Fang, G. Guan, J. Ren, H. Peng, A gum-like lithium-ion battery based on a novel arched structure, Adv. Mater. 27 (2015) 1363–1369. https://doi.org/10.1002/adma.201405127

[5]	D. Larcher, J.-M. Tarascon, Towards greener and more sustainable batteries for electrical energy storage, Nature Chem. 7 (2015) 19–29. https://doi.org/10.1038/nchem.2085

[6]	Y. Ding, G. Yu, A Bio-Inspired, Heavy-Metal-Free, Dual-electrolyte liquid battery towards sustainable energy storage, Angew. Chem. Int. Ed. 55 (2016) 4772–4776. https://doi.org/10.1002/anie.201600705

[7]	K. Jost, D.P. Durkin, L.M. Haverhals, E. Kathryn Brown, M. Langenstein, H.C. De Long, P.C. Trulove, Y. Gogotsi, G. Dion, K. Jost, M. Langenstein, Y.A. Gogotsi J Drexel, G. Dion, D.P. Durkin, E.K. Brown, P.C. Trulove, L.M. Haverhals, H.C. De Long, Natural fiber welded electrode yarns for knittable textile supercapacitors, Adv. Energy Mater. 5 (2015) 1401286. https://doi.org/10.1002/aenm.201401286

[8]	J.-H. Lee, J.H. Lee, Y.J. Lee, K.T. Nam, Protein/peptide based nanomaterials for energy application, Curr. Opin. Biotechnol. 24 (2013) 599–605. https://doi.org/10.1016/j.copbio.2013.02.004

[9]	P. Hu, H. Wang, Y. Yang, J. Yang, J. Lin, L. Guo, Renewable-biomolecule-based full lithium-ion batteries, Adv. Mater. 28 (2016) 3486–3492. https://doi.org/10.1002/adma.201505917

[10]　P. Poizot, F. Dolhem, Clean energy new deal for a sustainable world: from non-CO_2 generating energy sources to greener electrochemical storage devices, Energy Environ. Sci. 4 (2011) 2003. https://doi.org/10.1039/c0ee00731e

[11]　Y. Yang, H. Wang, R. Hao, L. Guo, Transition-metal-free biomolecule-based flexible asymmetric supercapacitors, Small 12 (2016) 4683–4689. https://doi.org/10.1002/smll.201503924

[12]　K. Mensah-darkwa, C. Zequine, P.K. Kahol, R.K. Gupta, Supercapacitor energy storage device using biowastes : A sustainable approach to green energy, Sustainability 11 (2019) 414. https://doi.org/10.3390/su11020414

[13]　Y. Zhai, Y. Dou, D. Zhao, P.F. Fulvio, R.T. Mayes, S. Dai, Carbon materials for chemical capacitive energy storage, Adv. Mater. 23 (2011) 4828–4850. https://doi.org/10.1002/adma.201100984

[14]　F. Béguin, V. Presser, A. Balducci, E. Frackowiak, Supercapacitors: Carbons and electrolytes for advanced supercapacitors, Adv. Mater. 26 (2014) 2283–2283. https://doi.org/10.1002/adma.201470093

[15]　X. Zhao, B.M. Sánchez, P.J. Dobson, P.S. Grant, The role of nanomaterials in redox-based supercapacitors for next generation energy storage devices, Nanoscale 3 (2011) 839. https://doi.org/10.1039/c0nr00594k

[16]　D.P. Dubal, P. Gomez-Romero, All nanocarbon Li-Ion capacitor with high energy and high power density, Mater. Today Energy 8 (2018) 109–117. https://doi.org/10.1016/j.mtener.2018.03.005

[17]　S.N. Beznosov, P.S. Veluri, M.G. Pyatibratov, A. Chatterjee, D.R. MacFarlane, O. V Fedorov, S. Mitra, Flagellar filament bio-templated inorganic oxide materials - towards an efficient lithium battery anode, Sci. Rep. 5 (2015) 7736. https://doi.org/10.1038/srep07736

[18]　S. De, A.M. Balu, J.C. van der Waal, R. Luque, Biomass-derived porous carbon materials: synthesis and catalytic applications, ChemCatChem. 7 (2015) 1608–1629. https://doi.org/10.1002/cctc.201500081

[19]　S. Jung, Y. Myung, B.N. Kim, I.G. Kim, I.-K. You, T. Kim, Activated biomass-derived graphene-based carbons for supercapacitors with high energy and power density, Sci. Rep. 8 (2018) 1915. https://doi.org/10.1038/s41598-018-20096-8

[20]　Y. Liu, J. Chen, B. Cui, P. Yin, C. Zhang, Design and Preparation of Biomass-Derived Carbon Materials for Supercapacitors: A Review, C. J. Carbon Res. 4 (2018) 53. https://doi.org/10.3390/c4040053

[21]　P. Simon, Y. Gogotsi, Charge storage mechanism in nanoporous carbons and its consequence for electrical double layer capacitors, Philos. Trans. R. Soc. A Math. Phys. Eng. Sci. 368 (2010) 3457–3467. https://doi.org/10.1098/rsta.2010.0109

[22]　A. Mahmoud, J. Olivier, J. Vaxelaire, A.F.A. Hoadley, Electrical field: A historical review of its application and contributions in wastewater sludge dewatering, Water Res. 44 (2010) 2381–2407. https://doi.org/10.1016/j.watres.2010.01.033

[23]　W. Qian, F. Sun, Y. Xu, L. Qiu, C. Liu, S. Wang, F. Yan, Human hair-derived carbon flakes for electrochemical supercapacitors, Energy Environ. Sci. 7 (2014) 379–386. https://doi.org/10.1039/C3EE43111H

[24]　C. Chen, D. Yu, G. Zhao, B. Du, W. Tang, L. Sun, Y. Sun, F. Besenbacher, M. Yu, Three-dimensional scaffolding framework of porous carbon nanosheets derived from plant wastes for high-performance supercapacitors, Nano Energy 27 (2016) 377–389. https://doi.org/10.1016/j.nanoen.2016.07.020

[25]　C.K. Ranaweera, P.K. Kahol, M. Ghimire, S.R. Mishra, R.K. Gupta, C.K. Ranaweera, P.K. Kahol, M. Ghimire, S.R. Mishra, R.K. Gupta, Orange-peel-derived carbon: designing sustainable and high-performance supercapacitor electrodes, C. J. Carbon Res. 3 (2017) 25. https://doi.org/10.3390/c3030025

[26]　Y.Q. Dang, S.Z. Ren, G. Liu, J. Cai, Y. Zhang, J. Qiu, Y.Q. Dang, S.Z. Ren, G. Liu, J. Cai, Y. Zhang, J. Qiu, Electrochemical and capacitive properties of carbon dots/reduced graphene oxide supercapacitors, Nanomaterials 6 (2016) 212. https://doi.org/10.3390/nano6110212

[27]　J.R. McDonough, J.W. Choi, Y. Yang, F. La Mantia, Y. Zhang, Y. Cui, Carbon nanofiber supercapacitors with large areal capacitances, Appl. Phys. Lett. 95 (2009) 243109. https://doi.org/10.1063/1.3273864

[28]　J. Yoon, J. Lee, J. Hur, J. Yoon, J. Lee, J. Hur, Stretchable Supercapacitors based on carbon nanotubes-deposited rubber polymer nanofibers electrodes with high tolerance against strain, Nanomaterials 8 (2018) 541. https://doi.org/10.3390/nano8070541

[29]　Z. Ling, Z. Wang, M. Zhang, C. Yu, G. Wang, Y. Dong, S. Liu, Y. Wang, J. Qiu, Sustainable Synthesis and Assembly of Biomass-derived B/N Co-doped carbon nanosheets with ultrahigh aspect ratio for high-performance supercapacitors, Adv. Funct. Mater. 26 (2016) 111–119. https://doi.org/10.1002/adfm.201504004

[30]　Z. Li, L. Zhang, B.S. Amirkhiz, X. Tan, Z. Xu, H. Wang, B.C. Olsen, C.M.B. Holt, D. Mitlin, Carbonized Chicken Eggshell membranes with 3D architectures as

high-performance electrode materials for supercapacitors, Adv. Energy Mater. 2 (2012) 431–437. https://doi.org/10.1002/aenm.201100548

[31] B. Duan, X. Gao, X. Yao, Y. Fang, L. Huang, J. Zhou, L. Zhang, Unique elastic N-doped carbon nanofibrous microspheres with hierarchical porosity derived from renewable chitin for high rate supercapacitors, Nano Energy 27 (2016) 482–491. https://doi.org/10.1016/j.nanoen.2016.07.034

[32] C. Long, X. Chen, L. Jiang, L. Zhi, Z. Fan, Porous layer-stacking carbon derived from in-built template in biomass for high volumetric performance supercapacitors, Nano Energy 12 (2015) 141–151. https://doi.org/10.1016/j.nanoen.2014.12.014

[33] S. Gao, X. Li, L. Li, X. Wei, A versatile biomass derived carbon material for oxygen reduction reaction, supercapacitors and oil/water separation, Nano Energy 33 (2017) 334–342. https://doi.org/10.1016/j.nanoen.2017.01.045

[34] Y. Li, G. Wang, T. Wei, Z. Fan, P. Yan, Nitrogen and sulfur co-doped porous carbon nanosheets derived from willow catkin for supercapacitors, Nano Energy 19 (2016) 165–175. https://doi.org/10.1016/j.nanoen.2015.10.038

[35] B. Li, F. Dai, Q. Xiao, L. Yang, J. Shen, C. Zhang, M. Cai, Activated carbon from biomass transfer for high-energy density lithium-ion supercapacitors, Adv. Energy Mater. 6 (2016) 1600802. https://doi.org/10.1002/aenm.201600802

[36] L. Zhu, F. Shen, R.L. Smith, L. Yan, L. Li, X. Qi, Black liquor-derived porous carbons from rice straw for high-performance supercapacitors, Chem. Eng. J. 316 (2017) 770–777. https://doi.org/10.1016/j.cej.2017.02.034

[37] W. Liu, J. Mei, G. Liu, Q. Kou, T. Yi, S. Xiao, Nitrogen-doped hierarchical porous carbon from wheat straw for supercapacitors, ACS Sustain. Chem. Eng. 6 (2018) 11595–11605. https://doi.org/10.1021/acssuschemeng.8b01798

[38] X.L. Su, M.Y. Cheng, L. Fu, J.H. Yang, X.C. Zheng, X.X. Guan, Superior supercapacitive performance of hollow activated carbon nanomesh with hierarchical structure derived from poplar catkins, J. Power Sources 362 (2017) 27–38. https://doi.org/10.1016/j.jpowsour.2017.07.021

[39] C.C. Hu, C.C. Wang, F.C. Wu, R.L. Tseng, Characterization of pistachio shell-derived carbons activated by a combination of KOH and CO2 for electric double-layer capacitors, Electrochim. Acta 52 (2007) 2498–2505. https://doi.org/10.1016/j.electacta.2006.08.061

[40] C. Dai, J. Wan, W. Geng, S. Song, F. Ma, J. Shao, KOH direct treatment of kombucha and in situ activation to prepare hierarchical porous carbon for high-

performance supercapacitor electrodes, J. Solid State Electrochem. 21 (2017) 2929–2938. https://doi.org/10.1007/s10008-017-3631-2

[41]   F. Li, F. Qin, K. Zhang, J. Fang, Y. Lai, J. Li, Hierarchically porous carbon derived from banana peel for lithium sulfur battery with high areal and gravimetric sulfur loading, J. Power Sources 362 (2017) 160–167. https://doi.org/10.1016/j.jpowsour.2017.07.038

[42]   H. Zhu, J. Yin, X. Wang, H. Wang, X. Yang, Microorganism-derived heteroatom-doped carbon materials for oxygen reduction and supercapacitors, Adv. Funct. Mater. 23 (2013) 1305–1312. https://doi.org/10.1002/adfm.201201643

[43]   N. Sudhan, K. Subramani, M. Karnan, N. Ilayaraja, M. Sathish, Biomass-derived activated porous carbon from rice straw for a high-energy symmetric supercapacitor in aqueous and non-aqueous electrolytes, Energy & Fuels 31 (2017) 977–985. https://doi.org/10.1021/acs.energyfuels.6b01829

[44]   M. Karnan, K. Subramani, P.K. Srividhya, M. Sathish, Electrochemical studies on corncob derived activated porous carbon for supercapacitors application in aqueous and non-aqueous electrolytes, Electrochim. Acta 228 (2017) 586–596. https://doi.org/10.1016/j.electacta.2017.01.095

[45]   J. McDonald-Wharry, M. Manley-Harris, K. Pickering, A comparison of the charring and carbonisation of oxygen-rich precursors with the thermal reduction of graphene oxide, Philos. Mag. 95 (2015) 4054–4077. https://doi.org/10.1080/14786435.2015.1108525

[46]   R. Hao, H. Lan, C. Kuang, H. Wang, L. Guo, Superior potassium storage in chitin-derived natural nitrogen-doped carbon nanofibers, Carbon 128 (2018) 224–230. https://doi.org/10.1016/j.carbon.2017.11.064

[47]   J. Bedia, M. Peñas-Garzón, A. Goméz-Avilés, J.J. Rodríguez, C. Belver, A Review on synthesis and characterization of biomass-derived carbons for adsorption of emerging contaminants from water, C. 4 (2018) 63. https://doi.org/10.3390/c4040063

[48]   N.C. Rath, S. Makkar, B. Packialakshmi, A.M. Donoghue, Egg shell membrane improves immunity of post hatch poultry: a paradigm for nutritional immunomodulation, n.d. https://www.ars.usda.gov/alternativestoantibiotics/Symposium2016/includes/Oral Presentations/4 - Immune Related Products/pdfs/3 RathATA symposiumFinal.pdf (accessed February 12, 2019).

[49]   T. Yang, T. Qian, M. Wang, X. Shen, N. Xu, Z. Sun, C. Yan, A sustainable route from biomass byproduct okara to high content nitrogen-doped carbon sheets for efficient sodium ion batteries, Adv. Mater. 28 (2016) 539–545. https://doi.org/10.1002/adma.201503221

[50]   R. Berenguer, F.J. García-Mateos, R. Ruiz-Rosas, D. Cazorla-Amorós, E. Morallón, J. Rodríguez-Mirasol, T. Cordero, Biomass-derived binderless fibrous carbon electrodes for ultrafast energy storage, Green Chem. 18 (2016) 1506–1515. https://doi.org/10.1039/C5GC02409A

[51]   B. Cagnon, X. Py, A. Guillot, F. Stoeckli, G. Chambat, Contributions of hemicellulose, cellulose and lignin to the mass and the porous properties of chars and steam activated carbons from various lignocellulosic precursors, Bioresour. Technol. 100 (2009) 292–298. https://doi.org/10.1016/j.biortech.2008.06.009

[52]   D.W. Rutherford, R.L. Wershaw, L.G. Cox, Changes in composition and porosity occurring during the thermal degradation of wood and wood components, Reston, VA, USA, 2004. http://www.usgs.gov/ (accessed February 13, 2019). https://doi.org/10.3133/sir20045292

[53]   X. Cao, K.S. Ro, J.A. Libra, C.I. Kammann, I. Lima, N. Berge, L. Li, Y. Li, N. Chen, J. Yang, B. Deng, J. Mao, Effects of biomass types and carbonization conditions on the chemical characteristics of hydrochars, J. Agric. Food Chem. 61 (2013) 9401–9411. https://doi.org/10.1021/jf402345k

[54]   J. Zhang, Y.S. Choi, C.G. Yoo, T.H. Kim, R.C. Brown, B.H. Shanks, Cellulose–hemicellulose and cellulose–lignin interactions during fast pyrolysis, ACS Sustain. Chem. Eng. 3 (2015) 293–301. https://doi.org/10.1021/sc500664h

[55]   K. Yu, J. Wang, K. Song, X. Wang, C. Liang, Y. Dou, K. Yu, J. Wang, K. Song, X. Wang, C. Liang, Y. Dou, Hydrothermal synthesis of cellulose-derived carbon nanospheres from corn straw as anode materials for lithium ion batteries, Nanomaterials 9 (2019) 93. https://doi.org/10.3390/nano9010093

[56]   D.R. Morais, E.M. Rotta, S.C. Sargi, E.G. Bonafe, R.M. Suzuki, N.E. Souza, M. Matsushita, J. V Visentainer, Proximate composition, mineral contents and fatty acid composition of the different parts and dried peels of tropical fruits cultivated in brazil, Artic. J. Braz. Chem. Soc. 28 (2017) 308–318. https://doi.org/10.5935/0103-5053.20160178

[57]   F. Dibanda Romelle, R.P. Ashwini, R.S. Manohar, Chemical composition of some selected fruit peels, European Am. J. (2016) 12-21.

Biomass Based Energy Storage Materials   Materials Research Forum LLC
Materials Research Foundations **78** (2020) 21-49   https://doi.org/10.21741/9781644900871-2

[58] R. Sánchez Orozco, P. Balderas Hernández, G. Roa Morales, F. Ureña Núñez, J. Orozco Villafuerte, V. Lugo Lugo, N. Flores Ramírez, C.E. Barrera Díaz, P. Cajero Vázquez, Characterization of lignocellulosic fruit waste as an alternative feedstock for bioethanol production, Bio. Resources. 9 (2014) 1873–1885. https://doi.org/10.15376/biores.9.2.1873-1885

[59] X.M. Wang, J. Zhang, L.H. Wu, Y.L. Zhao, T. Li, J.Q. Li, Y.Z. Wang, H.-G. Liu, A mini-review of chemical composition and nutritional value of edible wild-grown mushroom from China, Food Chem. 151 (2014) 279–285. https://doi.org/10.1016/j.foodchem.2013.11.062

[60] J. Arroyo, V. Farkaš, A.B. Sanz, E. Cabib, 'Strengthening the fungal cell wall through chitin-glucan cross-links: effects on morphogenesis and cell integrity, Cell. Microbiol. 18 (2016) 1239–1250. https://doi.org/10.1111/cmi.12615

[61] M. Chen, L. Wang, J. Hou, S. Yang, X. Zheng, L. Chen, X. Li, M. Chen, L. Wang, J. Hou, S. Yang, X. Zheng, L. Chen, X. Li, Mycoextraction: Rapid cadmium removal by macrofungi-based technology from alkaline soil, Minerals 8 (2018) 589. https://doi.org/10.3390/min8120589

[62] C. Sales-Campos, L.M. Araujo, M.T. de A. Minhoni, M.C.N. de Andrade, Physiochemical analysis and centesimal composition of Pleurotus ostreatus mushroom grown in residues from the Amazon, Ciência e Tecnol. Aliment 31 (2011) 456–461. https://doi.org/10.1590/S0101-20612011000200027

[63] W. Arbia, L. Arbia, L. Adour, A. Amrane, Chitin extraction from crustacean shells using biological methods – A review, Food Technol. Biotechnol. 51 (2012) 12–25.

[64] J. Majtán, K. Bíliková, O. Markovič, J. Gróf, G. Kogan, J. Šimúth, Isolation and characterization of chitin from bumblebee (Bombus terrestris), Int. J. Biol. Macromol. 40 (2007) 237–241. https://doi.org/10.1016/j.ijbiomac.2006.07.010

[65] M. Kaya, E. Lelešius, R. Nagrockaitė, I. Sargin, G. Arslan, A. Mol, T. Baran, E. Can, B. Bitim, Differentiations of chitin content and surface morphologies of chitins extracted from male and female grasshopper species, PLoS One 10 (2015) e0115531. https://doi.org/10.1371/journal.pone.0115531

[66] E. Kovaleva, A. Pestov, D. Stepanova, L. Molochnikov, Characterization of chitin and its complexes extracted from natural raw sources, in: AIP Conf. Proc., AIP Publishing LLC, 2016: p. 050007. https://doi.org/10.1063/1.4964577

[67] J.-P. Latgé, The cell wall: a carbohydrate armour for the fungal cell, Mol. Microbiol. 66 (2007) 279–290. https://doi.org/10.1111/j.1365-2958.2007.05872.x

[68] H. Schwarz, B. Moussian, Electron-microscopic and genetic dissection of arthropod cuticle differentiation, Modern Research and Educational Topics in Microscopy (227) 316-325.

[69] J. McKittrick, P.-Y. Chen, S.G. Bodde, W. Yang, E.E. Novitskaya, M.A. Meyers, The structure, functions, and mechanical properties of keratin, JOM 64 (2012) 449–468. https://doi.org/10.1007/s11837-012-0302-8

[70] C.R. Robbins, Chemical composition of different hair types, in: Chem. Phys. Behav. Hum. Hair, Springer Berlin Heidelberg, Berlin, Heidelberg, 2012: pp. 105–176. https://doi.org/10.1007/978-3-642-25611-0_2

[71] Y. Lv, F. Zhang, Y. Dou, Y. Zhai, J. Wang, H. Liu, Y. Xia, B. Tu, D. Zhao, A comprehensive study on KOH activation of ordered mesoporous carbons and their supercapacitor application, J. Mater. Chem. 22 (2012) 93–99. https://doi.org/10.1039/C1JM12742J

[72] J. Wang, S. Kaskel, KOH activation of carbon-based materials for energy storage, J. Mater. Chem. 22 (2012) 23710. https://doi.org/10.1039/c2jm34066f

[73] J. Ajuria, E. Redondo, M. Arnaiz, R. Mysyk, T. Rojo, E. Goikolea, Lithium and sodium ion capacitors with high energy and power densities based on carbons from recycled olive pits, J. Power Sources 359 (2017) 17–26. https://doi.org/10.1016/j.jpowsour.2017.04.107

[74] H. Zhu, X. Wang, F. Yang, X. Yang, Promising carbons for supercapacitors derived from fungi, Adv. Mater. 23 (2011) 2745–2748. https://doi.org/10.1002/adma.201100901

[75] W. Liu, J. Mei, G. Liu, Q. Kou, T. Yi, S. Xiao, Nitrogen-doped hierarchical porous carbon from wheat straw for supercapacitors, ACS Sustain. Chem. Eng. 6 (2018) 11595–11605. https://doi.org/10.1021/acssuschemeng.8b01798

[76] S. Yang, K. Zhang, S. Yang, K. Zhang, Converting corncob to activated porous carbon for supercapacitor application, Nanomaterials 8 (2018) 181. https://doi.org/10.3390/nano8040181

[77] Y. Liu, X. Zhang, S. Poyraz, C. Zhang, J.H. Xin, One-step synthesis of multifunctional zinc-iron-oxide hybrid carbon nanowires by chemical fusion for supercapacitors and interfacial water marbles, ChemNanoMat 4 (2018) 546–556. https://doi.org/10.1002/cnma.201800075

[78] S.N. Talapaneni, J.H. Lee, S.H. Je, O. Buyukcakir, T. Kwon, K. Polychronopoulou, J.W. Choi, A. Coskun, Chemical blowing approach for

ultramicroporous carbon nitride frameworks and their applications in gas and energy storage, Adv. Funct. Mater. 27 (2017) 1604658. https://doi.org/10.1002/adfm.201604658

[79]   L. Yin, Y. Chen, D. Li, X. Zhao, B. Hou, B. Cao, 3-Dimensional hierarchical porous activated carbon derived from coconut fibers with high-rate performance for symmetric supercapacitors, Mater. Des. 111 (2016) 44–50. https://doi.org/10.1016/j.matdes.2016.08.070

[80]   W.-H. Qu, Y.-Y. Xu, A.-H. Lu, X.-Q. Zhang, W.-C. Li, Converting biowaste corncob residue into high value added porous carbon for supercapacitor electrodes, Bioresour. Technol. 189 (2015) 285–291. https://doi.org/10.1016/j.biortech.2015.04.005

[81]   P. Yu, Z. Zhang, L. Zheng, F. Teng, L. Hu, X. Fang, A novel sustainable flour derived hierarchical nitrogen-doped porous carbon/polyaniline electrode for advanced asymmetric supercapacitors, Adv. Energy Mater. 6 (2016) 1601111. https://doi.org/10.1002/aenm.201601111

[82]   Y. Gong, D. Li, C. Luo, Q. Fu, C. Pan, Highly porous graphitic biomass carbon as advanced electrode materials for supercapacitors, Green Chem. 19 (2017) 4132–4140. https://doi.org/10.1039/C7GC01681F

[83]   M. Fu, W. Chen, X. Zhu, B. Yang, Q. Liu, Crab shell derived multi-hierarchical carbon materials as a typical recycling of waste for high performance supercapacitors, Carbon 141 (2019) 748–757. https://doi.org/10.1016/j.carbon.2018.10.034

[84]   X.L. Su, S.H. Li, S. Jiang, Z.K. Peng, X.X. Guan, X.C. Zheng, Superior capacitive behavior of porous activated carbon tubes derived from biomass waste-cotonier strobili fibers, Adv. Powder Technol. 29 (2018) 2097–2107. https://doi.org/10.1016/j.apt.2018.05.018

[85]   M. Zhou, J. Gomez, B. Li, Y.B. Jiang, S. Deng, Oil tea shell derived porous carbon with an extremely large specific surface area and modification with MnO_2 for high-performance supercapacitor electrodes, Appl. Mater. Today 7 (2017) 47–54. https://doi.org/10.1016/j.apmt.2017.01.008

[86]   H. Ba, W. Wang, S. Pronkin, T. Romero, W. Baaziz, L. Nguyen-Dinh, W. Chu, O. Ersen, C. Pham-Huu, Biosourced foam-like activated carbon materials as high-performance supercapacitors, Adv. Sustain. Syst. 2 (2018) 1700123. https://doi.org/10.1002/adsu.201700123

[87]   B. Liu, L. Zhang, P. Qi, M. Zhu, G. Wang, Y. Ma, X. Guo, H. Chen, B. Zhang, Z. Zhao, B. Dai, F. Yu, Nitrogen-doped banana peel–derived porous carbon foam as binder-free electrode for supercapacitors, Nanomaterials 6 (2016) 18. https://doi.org/10.3390/nano6010018

[88]   K. Subramani, N. Sudhan, M. Karnan, M. Sathish, Orange peel derived activated carbon for fabrication of high-energy and high-rate supercapacitors, Chemistry Select 2 (2017) 11384–11392. https://doi.org/10.1002/slct.201701857

[89]   Q. Yao, H. Wang, C. Wang, C. Jin, Q. Sun, One step construction of nitrogen–carbon derived from bradyrhizobium japonicum for supercapacitor applications with a soybean leaf as a separator, ACS Sustain. Chem. Eng. 6 (2018) 4695–4704. https://doi.org/10.1021/acssuschemeng.7b03777

[90]   Q. Zhang, K. Han, S. Li, M. Li, J. Li, K. Ren, Synthesis of garlic skin-derived 3D hierarchical porous carbon for high-performance supercapacitors, Nanoscale 10 (2018) 2427–2437. https://doi.org/10.1039/C7NR07158B

[91]   T. Kesavan, T. Partheeban, M. Vivekanantha, M. Kundu, G. Maduraiveeran, M. Sasidharan, Hierarchical nanoporous activated carbon as potential electrode materials for high performance electrochemical supercapacitor, Microporous Mesoporous Mater. 274 (2019) 236–244. https://doi.org/10.1016/j.micromeso.2018.08.006

[92]   S. Ahmed, A. Ahmed, M. Rafat, Nitrogen doped activated carbon from pea skin for high performance supercapacitor, Mater. Res. Express 5 (2018) 045508. https://doi.org/10.1088/2053-1591/aabbe7

[93]   A. Mahto, R. Gupta, K.K. Ghara, D.N. Srivastava, P. Maiti, K. D., P.Z. Rivera, R. Meena, S.K. Nataraj, Development of high-performance supercapacitor electrode derived from sugar industry spent wash waste, J. Hazard. Mater. 340 (2017) 189–201. https://doi.org/10.1016/j.jhazmat.2017.06.048

[94]   Y. Huang, L. Peng, Y. Liu, G. Zhao, J.Y. Chen, G. Yu, Biobased Nano Porous Active Carbon Fibers for High-Performance Supercapacitors, ACS Appl. Mater. Interfaces 8 (2016) 15205–15215. https://doi.org/10.1021/acsami.6b02214

[95]   N. Guo, M. Li, Y. Wang, X. Sun, F. Wang, R. Yang, Soybean root-derived hierarchical porous carbon as electrode material for high-performance supercapacitors in ionic liquids, ACS Appl. Mater. Interfaces 8 (2016) 33626–33634. https://doi.org/10.1021/acsami.6b11162

[96]   E.Y.L. Teo, L. Muniandy, E.P. Ng, F. Adam, A.R. Mohamed, R. Jose, K.F. Chong, High surface area activated carbon from rice husk as a high performance

supercapacitor electrode, Electrochim. Acta 192 (2016) 110–119.
https://doi.org/10.1016/j.electacta.2016.01.140

[97]　S.T. Senthilkumar, N. Fu, Y. Liu, Y. Wang, L. Zhou, H. Huang, Flexible fiber hybrid supercapacitor with $NiCo_2O_4$ nanograss@carbon fiber and bio-waste derived high surface area porous carbon, Electrochim. Acta 211 (2016) 411–419. https://doi.org/10.1016/j.electacta.2016.06.059

[98]　A.K. Mondal, K. Kretschmer, Y. Zhao, H. Liu, C. Wang, B. Sun, G. Wang, Nitrogen-doped porous carbon nanosheets from eco-friendly eucalyptus leaves as high performance electrode materials for supercapacitors and lithium ion batteries, Chem. A Eur. J. 23 (2017) 3683–3690. https://doi.org/10.1002/chem.201605019

[99]　H. Ba, W. Wang, S. Pronkin, T. Romero, W. Baaziz, L. Nguyen-Dinh, W. Chu, O. Ersen, C. Pham-Huu, Biosourced foam-like activated carbon materials as high-performance supercapacitors, Adv. Sustain. Syst. 2 (2018) 1700123. https://doi.org/10.1002/adsu.201700123

[100]　P. Yu, Z. Zhang, L. Zheng, F. Teng, L. Hu, X. Fang, A novel sustainable flour derived hierarchical nitrogen-doped porous carbon/polyaniline electrode for advanced asymmetric supercapacitors, Adv. Energy Mater. 6 (2016) 1601111. https://doi.org/10.1002/aenm.201601111

[101]　R. Fang, P. Tian, X. Yang, R. Luque, Y. Li, Encapsulation of ultrafine metal-oxide nanoparticles within mesopores for biomass-derived catalytic applications, Chem. Sci. 9 (2018) 1854–1859. https://doi.org/10.1039/C7SC04724J

[102]　J. Wei, Y. Liang, Y. Hu, B. Kong, G.P. Simon, J. Zhang, S.P. Jiang, H. Wang, A versatile iron-tannin-framework ink coating strategy to fabricate biomass-derived iron carbide/fe-n-carbon catalysts for efficient oxygen reduction, Angew. Chem. Int. Ed. 55 (2016) 1355–1359. https://doi.org/10.1002/anie.201509024

[103]　W.S. Cha, S.N. Talapaneni, D.M. Kempaiah, S. Joseph, K.S. Lakhi, A.M. Al-Enizi, D.H. Park, A. Vinu, Excellent supercapacitance performance of 3-D mesoporous carbon with large pores from FDU-12 prepared using a microwave method, RSC Adv. 8 (2018) 17017–17024. https://doi.org/10.1039/C8RA01281D

[104]　Z. Ling, Z. Wang, M. Zhang, C. Yu, G. Wang, Y. Dong, S. Liu, Y. Wang, J. Qiu, Sustainable synthesis and assembly of biomass-derived B/N co-doped carbon nanosheets with ultrahigh aspect ratio for high-performance supercapacitors, Adv. Funct. Mater. 26 (2016) 111-119. https://doi.org/10.1002/adfm.201504004

Materials Research Forum LLC

https://doi.org/10.21741/9781644900871-3

Chapter 3

Biomass Derived Composites for Energy Storage

Pitchaimani Veerakumar[1,2,*] and King-Chuen Lin[1,2,*]

[1]Department of Chemistry, National Taiwan University, Taipei 10617, Taiwan

[2]Institute of Atomic and Molecular Sciences, Academia Sinica, Taipei 10617, Taiwan

*Email (spveerakumar@gmail.com (P.V.); kclin@ntu.edu.tw (K.-C. L.)

Abstract

Recently, biomass-derived valuable carbon structures have become a research focus that offer multiple contributions to electrochemical energy storage. This book chapter reveals the nanostructured materials synthesized from plant-based biomass in the design of high-performance superconductors. Encountering of challenges and limitations for better achievement have been discussed from the point-of-view of materials preparation. The aim of this chapter is to offer an overview of the prior studies of the common types of biomass-related carbons, their potential advantages in supercapacitor applications, and the relationship of electrochemical properties of biomass-derived carbon materials and their structures.

Keywords

Biomass, Porous Carbons, Supercapacitors, Energy Storage, Energy Density, Power Density

Contents

1. **Introduction**...**51**

2. **Sustainable biomass-carbon materials**...**54**

3. **Calculation paramaters**..**55**

4. **Biomass activation**...**56**

    4.1   Physical activation..**58**

    4.2   Chemical activation...**60**

    4.3   Hydrothermal carbonization...**64**

Biomass Based Energy Storage Materials   Materials Research Forum LLC
Materials Research Foundations **78** (2020) 50-90   https://doi.org/10.21741/9781644900871-3

4.4   Other activations 67

5.   Outlook **74**

Conclusions and prospects **75**

References **76**

1.   Introduction

Porous carbons (PCs) have attracted considerable interest in their utilization as relatively low cost, non-toxicity, and environment-friendly materials. PCs have been explored in important applications as catalysts, CO_2 capture, electrochemical sensor, energy storage, fuel cell, solar cells, and so on [1–4]. The electrochemical capacitor (EC) also named as supercapacitor (SC) is a novel type of energy storage device whose capacitance is 20–200 times higher than that of an electrolytic capacitor [5]. The charge storage mechanisms of ECs have been divided into two categories. One is the electric double-layer capacitor (EDLC) and the other is the Faradic pseudo-capacitance capacitor [6]. In general, SC devices are composed of an anode and a cathode in a cell with electrolyte and a separator separating the two electrodes. Depending on the electrode materials, there are three types of SC devices containing symmetric supercapacitor (SSC), asymmetric supercapacitor (ASC), and hybrid supercapacitor (HSC), of which ASC and HSC easily cause confusion related to one another [7]. SSCs typically adopt activated carbon (AC) and pseudo-capacitive materials as two identical supercapacitor-type electrodes. Most commercial SCs rely on two symmetric AC electrodes in organic electrolytes with voltage window up to 2.7 V [8]. The resulting fast surface redox reactions are mainly caused by the charge storage of ion adsorption and pseudo-capacitors. The energy storage capacities of conventional dielectric capacitors are lower by several orders of magnitude than those of SCs. Thus, for their high power density (P), long cyclic stability, and high safety, the SCs can be applied as a complement to rechargeable batteries.

ASCs comprise two different supercapacitor-type electrodes, made of a double-layer carbon material and a pseudocapacitance material. For instance, as one promising ASC, AC/MnO_2 has been widely studied for energy storage [9]. In contrast, the HSCs contain a battery-type and supercapacitor-type electrodes [10–14]. Mostly, carbons are in three forms: diamond (crystalline), graphite (crystalline), and amorphous carbons. Several carbon-based materials, including carbon nanotubes (CNTs) [15], activated carbons (ACs) [16], carbon black (CB) [17], ordered mesoporous carbons (OMCs) [18], carbon fibers (CFs) [19], carbon aerogels (CAs) [20], glassy carbon, carbon whiskers and graphene [21,22] have been explored for potential applications in the field of materials science and chemistry (Fig. 1).

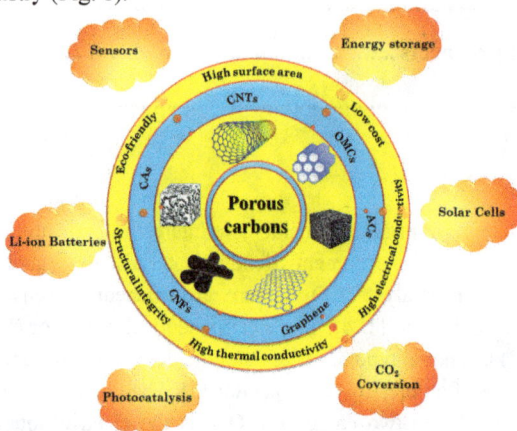

Figure 1. Porous carbon materials and their applications.

Metal oxides (RuO_2, ReO_x, NiO, MnO_2), conducting polymers, or N, O, S, Cl co-doped materials, while decorated on the surface of these carbons, can further enhance the carbons electrochemical performance [23–27]. Indeed, transition metal oxide-based nanocomposite is an important electrode material especially applied in SCs and lithium-ion batteries (LIBs) [28,29]. Nevertheless, NiO with high resistivity is not an appropriate electrode material applied to SC, of which the energy storage capability arises from the redox process at the electrode/electrolyte interface. To find out various NiO-containing composites through controlled chemical methods can enhance the NiO-based electrode conductivity to improve the SC electrochemical performance [30,31]. For instance, after conducting support by carbons, the NiO/C composite electrode has exhibited high specific capacitance (C_{sp}) and good power characteristics [32]. Since the past years, nanostructured carbons have been developed as electrode materials to increase the outer surface area and shorten the diffusion length for improving the performance of SCs [33–35]. Their energy/power performance depends on the charge storage mechanism. Comparison of charge-discharge capability of conventional capacitors, supercapacitors, and batteries is displayed in Fig. 2.

This chapter reports the latest progress on synthesis of electrode materials, supercapacitors in charge storage mechanisms, electrolytes, characterization, and applications. The newly developed charge storage mechanism for bridging the gap between conventional pseudocapacitor and battery is remarked. Further, discussion on the challenges and prospects regarding practical applications of supercapacitors is made. In this regard, various applications including hybrid, asymmetric, and symmetric SC have been extensively explored. As shown in Fig. 3, the recent developments of SC in charge storage systems and related materials for electrodes and electrolytes are displayed [7]. Herein, the features of the biomass-derived nanocomposites for energy storage SCs and their latest advances will be briefly summarized.

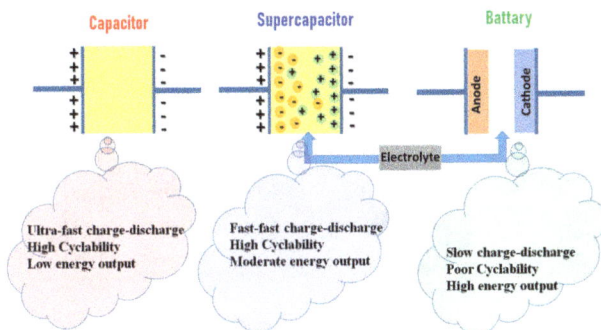

Figure 2. Characteristic outlines for conventional capacitors, supercapacitors and batteries.

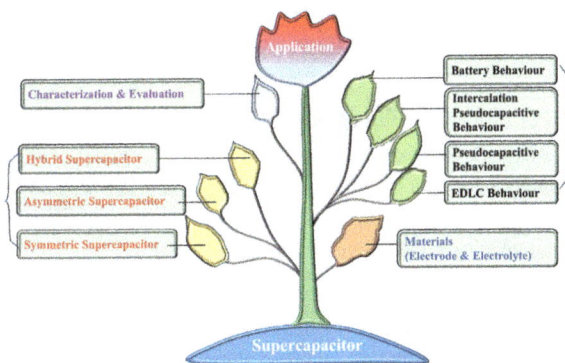

Figure 3. Schematic view of the various electrochemical capacitors [7]. Reproduced from ref [7]. Copyright©2016 The Royal Society of Chemistry.

2. Sustainable biomass-carbon materials

As a well-known raw material, biomass is used to produce carbon materials for SCs electrodes [36]. As a biological matter, biomass mainly consists of hemicelluloses $(C_5H_8O_4)_m$, cellulose $(C_6H_{10}O_5)_x$, lignin $[C_9H_{10}O_3(COH_3)_{0.9-1.7}]_n$, fats, proteins, arrowroots, sugars, water, ash, and small extractives. Among these, cellulose contains the largest fraction, 35–50% of biomass by weight [37], hemicelluloses 20-35%, and lignin 15-20%, while the remaining content is 15-20%. In an environmentally friendly way, these materials may be used to produce carbon materials as shown in Fig. 4a [38]. Biomass as a source to produce energy is available from agriculture, crops, and forests, (Fig. 4b) [39].

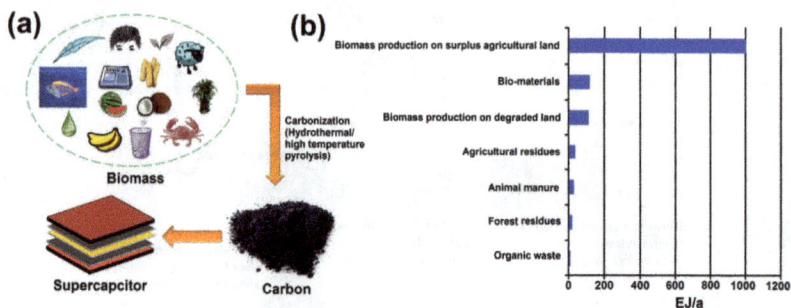

Figure 4. (a) Schematic for carbonization of biomass and its applications to fabricate supercapacitor cells for green energy storage [36]. Reproduced with permission from ref [36]. Copyright©2018 Springer Publications. and (b) biomass resources present for utilization [39]. Reproduced with permission from ref [39]. Copyright©2011 The American Chemical Society.

Various approaches for efficient utilization of biomass are possible. For instance, carbonization of biomass/waste offers an economically and green synthesis approach to provide carbon materials and recycle biomass. Carbon materials are potentially utilized in energy storage, sensing, and conversion devices including fuel cells, batteries, photovoltaic, supercapacitors and conventional capacitors [36].

Among the carbon materials, AC with amorphous feature has high porosity because of the production process and treatment [37]. The development of high surface area, chemical polarity and pore structure depends largely on the precursor material and activation process [40]. To compare with those commercially produced from fossil fuel-based precursor, biomass-based ACs are attracting the attention, because these are readily

Biomass Based Energy Storage Materials   Materials Research Forum LLC
Materials Research Foundations **78** (2020) 50-90   https://doi.org/10.21741/9781644900871-3

available, cheaper, structurally porous, renewable, and environmentally friendly [41]. The use of waste biomass, such as waste coffee beans [42], cassava peel waste [43], apricot shell [44], sugarcane bagasse [45], rice husk [46], peanut shell [47], sunflower seed shell [48], coffee endocarp [49], rubberwood sawdust [50], oil palm empty fruit bunch [51], and palm kernel shell [52], as precursors to produce porous carbons has becomes popular. Meanwhile, a great challenge is encountered since the direct discharge of some of these wastes may cause environmental problems.

3.   Calculation paramaters

In SC applications, the electrochemical parameters including C_{sp}, E and P have been related to each other [53,54]. The C_{sp} (c) from the CV measurements was calculated using the following Eq. 1:

$$C_{sp} = \frac{Q}{m \times \Delta V} \qquad (1)$$

where C_{sp} is the specific capacitance in units of F g^{-1}, Q is the average charge during the charge and discharge process (in units of C), m is the mass of the active material in the working electrode in units of g, and ΔV is the potential window in units of V. Equation (2) is applied in the case of galvanostatic charge/discharge (GCD) measurements. The discharge time is directly related to the C_{sp} of the three-electrode configuration as expressed by:

$$C_{sp} = \frac{I \times \Delta t}{m \times \Delta V} \qquad (2)$$

where I is the applied current density in units of A, Δt the charge/discharge duration in units of s.

In two electrodes symmetric cell configuration, C_{sp} derived from galvanostatic tests can be calculated from the Eq. 3:

$$C_{sp} = 2(\frac{I \times \Delta t}{m \times \Delta V}) \qquad (3)$$

where I is the constant discharge current, Δt is the discharging time, m is the mass of one electrode, and ΔV is the voltage drop upon discharging.

In contrast, in two electrodes *asymmetric* cell configuration, C_{sp} derived from galvanostatic tests can be calculated from the Eq. 4:

$$C_{sp} = \frac{I \times \Delta t}{m \times \Delta V} \qquad (4)$$

where I is the discharging current, Δt and ΔV *have the same notation as in Eq. 3*, and m is the *total mass* of the active electrode materials in the positive and negative electrode.

P and E of the sample derived from galvanostatic tests can be calculated from the following Eqs.5 and 6:

$$E\ (Wh/kg) = \frac{C_{sp} \times \Delta V^2}{7.2} \qquad (5)$$

$$P\ (W/kg) = \frac{E \times 3600}{t} \qquad (6)$$

where ΔV is the potential window (V), t is the discharge time (s), and C_{sp} of the electrode determined using charge–discharge measurements.

4. Biomass activation

Various carbonization treatments such as pyrolysis as well as hydrothermal carbonisation (HTC) and physical (also called thermal) and/or chemical activation methods can be used to convert biomass into carbon, as illustrated in Fig. 5. Physical activation applied to the AC products contains carbonization of a biomass precursor and gasification. The former treatment is to remove non-carbon species by thermal decomposition in an inert atmosphere, while the latter is to develop porosity by partial etching of carbon during the annealing with an oxidizing agent, such as CO_2, H_2O, or a stream [55,56]. The chemical activation method employs chemicals (such as acid, strong base, or salt) to increase the specific surface area (SSA). These chemicals include potassium hydroxide (KOH) [57], sodium hydroxide (NaOH) [58], potassium carbonate (K_2CO_3) [59], potassium phosphate dibasic (K_2HPO_4) [60], phosphoric acid (H_3PO_4) [61], zinc chloride ($ZnCl_2$) [62], sulfuric acid (H_2SO_4) [63], hydrochloric acid (HCl) [64], cobalt acetate ($Co(OAc)_2$) [65], hydrogen peroxide (H_2O_2), nitric acid (HNO_3), and ammonium peroxydisulfate (($NH_4)_2S_2O_8$) [66]. This process generally results in smaller pores, more uniform pore size distribution, SSA and (V_{sp}). The conversion of carbon at elevated temperatures is favourable for higher C_{sp} in both aqueous and organic electrolytes. Hence, these methods have been employed to transfer biomass into value-added carbon materials.

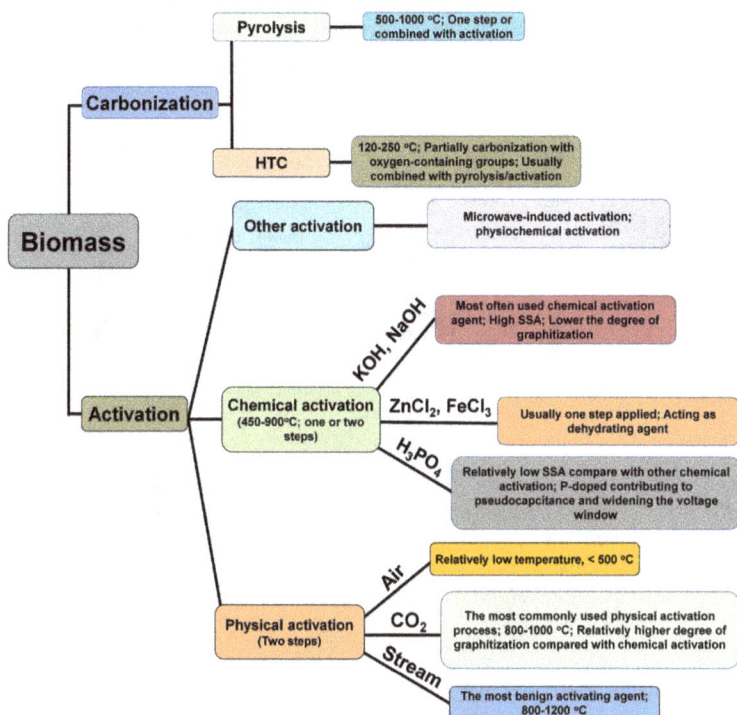

Figure 5. Common methods for converting biomass to carbon materials [55].
Reproduced with permission from ref [55]. Copyright©2017 The Royal Society of
Chemistry.

An interesting study [62] on PC produced from *Moringa oleifera* fruit shells containing cellulose, hemicellulose and lignin has shown that cellulose, hemicellulose, and aromatic regions tended to produce more curl-like porous carbons structures. With the assistance of $ZnCl_2$, a more environmentally friendly chemical activating agent than KOH, the authors propose a chemical activation method for biomass at 900 °C, where mixing $ZnCl_2$ in biomass in the ratio of 1:2, yielded a high surface area up to 2522 m^2 g^{-1} with co-existing micro/mesoporosity. As illustrated in Fig. 6, the Ru$_x$/MOC nanocomposites with varied Ru loadings (x= 1.0 and 1.5 wt%) were fabricated using a thermal reduction treatment at 900 °C, exhibiting excellent electrochemical properties with a maximum capacitance of 291 F g^{-1} at a current density of 1.0 A g^{-1} in 1.0 M H_2SO_4 electrolyte. While being facially fabricated by the eco-friendly and cost-effective route, these highly stable and durable Ru/MOC composites have potential applications in energy storage, biosensing, and catalysis.

Figure 6. The synthesis route for Ru/MOC nanocomposites. (a) Moringa Oleifera fruit shells; (b) microwave irradiation; (c) carbonization and activation; (d) Ru(acac)3 addition and then with MW reduction irradiation at 900 °C; (e) Ru/MOC-based electrode in application of supercapacitor [62]. Reproduced with permission from ref [62]. Copyright©2016 Nature publishing group.

On the other hand, there are two common activation approaches: physical activation using CO_2 while chemical activation using $ZnCl_2$. Both methods introduce additional gas (CO_2) or a chemical ($ZnCl_2$), from which the CO_2 releases during the activation process while the zinc compound is removed by using acid from the follow-up process which will cause environmental concerns.

4.1   Physical activation

As regarded to be time-consuming, energy-wasting, non-eco-friendly, and an intricate procedure, physical or thermal activation involves the processes (i) carbonization at low temperature (~700 to ~1150 K) and (ii) activation at high temperature (~900 to ~1200 K). Physical activation, mainly with CO_2 and H_2O steam treatment, is adopted to convert many biomass precursors into ACs for supercapacitors, including coffee endocarp, rubber wood sawdust, oil palm empty fruit bunches, paper flower, and pistachio shell [67]. Physical activation usually results in porous carbon with a bulk density and a high yield but with a small pore volume and a relatively low surface area and incapability to adjust the electronic and surface chemical properties of carbon materials. For this reason, Veerakumar et al. [68]. adopted both chemical ($ZnCl_2$) and physical activation (CO_2) methods to treat the carbon materials prepared from mature paper flower (PF). Such paper flower carbon (PFC) exhibited a high SSA of 1801 m^2 g^{-1} with a porous graphitic carbon layer structure, a large pore volume of 1.16 cm^3 g^{-1} and a C_{sp} of ~118 F g^{-1}. Additionally, the stability at 12 A g^{-1} was tested to reach 10,000 cycles with capacity retention of around 97.4%. The obtained porous carbon nanosheets can be used for both supercapacitor and dye removal applications (Fig 7).

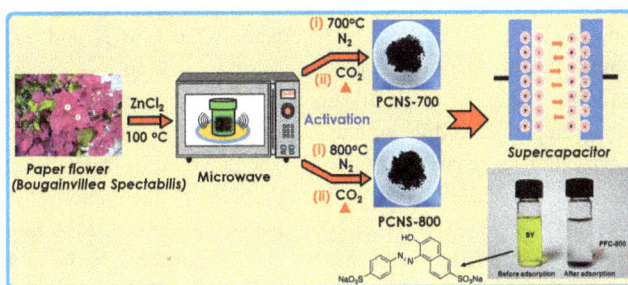

Figure 7. Schematic Illustration of the preparation route for PFC carbon for supercapacitor and dye removal applications [68]. Reproduced with permission from ref [68]. Copyright©2019 Elsevier publications.

Given high chemical stability, large surface area, scalable synthesis, reasonable cost, and biocompatibility, PFC has become promising in the device fabrication [69]. A redox reaction of AC proposed for pseudocapacitance in a 1.0 M H_2SO_4 electrolyte is illustrated below:

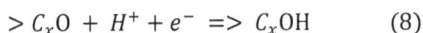

$$> C_x - O + H^+ => C_xO // H^+ \qquad (7)$$

$$> C_xO + H^+ + e^- => C_xOH \qquad (8)$$

where the symbol > denotes the surface of the PFC material, and the symbol // the adsorption interface between the oxygen functional groups and H^+ in the PFC. The reaction (8) indicates that the proton H^+ is adsorbed on the oxygen functional groups and also involved in the electron transfer [70].

4.2 Chemical activation

Chemical activation is performed with activating agents by a single step impregnation of precursors, followed by a heat treatment process at a moderate temperature ranging from ~700 to ~900 K. The chemical activating agents such as KOH/NaOH, K_2CO_3, $ZnCl_2$, and H_3PO_4 are frequently mixed with biomass precursors for creating porous structures [71]. However, in the activation process, a great amount of materials are often etched by the alkali metal, resulting in overall low yield. For extensive pores creation,-the collapse of carbon nanostructures and the loss of heteroatoms (such as N and S) may further happen to damage the electrochemical performance. Apart from KOH activation, $ZnCl_2$ activation similarly enhances the specific surface area and helps dehydrate the biomass precursor at low temperature while aromatic condensation at high temperature. But the non-etching activation mechanism limits the preparation of highly porous carbon materials with a BET surface area > 2000 m^2 g^{-1} [72]. The activation of biomass with KOH is used as an example. Its proceeding of the first step involved the carbonization of tree leaves into carbon at different temperatures in N_2 atmosphere. In the following step it was converted to porous high surface area carbon using KOH activation. It is a well-known process for the formation of pores on the carbon due to the intercalation of metallic potassium. Generally, such an intricate process involves several mechanisms proposed in the literature [60]. Based on the literature reports [73,74], the formation of porous network by KOH activation process has been explained via following eqns.

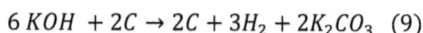

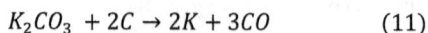

$$6\,KOH + 2C \rightarrow 2C + 3H_2 + 2K_2CO_3 \quad (9)$$

$$2K_2CO_3 \rightarrow K_2O + CO_2 \quad\quad\quad (10)$$

$$K_2CO_3 + 2C \rightarrow 2K + 3CO \quad\quad (11)$$

$$CO_2 + C \rightarrow 2CO \quad\quad\quad\quad\quad (12)$$

$$C + K_2O \rightarrow 2K + CO \quad\quad\quad (13)$$

$ZnCl_2$ acts as a dehydration agent during the carbonization process, resulting in charring and aromatization reactions of the carbon skeleton and development of the pore structure [75]. In addition, H_2 is evolved from the hydroaromatic structure of the precursor and leaves behind some sites for reaction. This treatment could form Lewis acid sites associated with Zn species (Zn-L). Zn-L promoted the dehydrogenation and

Biomass Based Energy Storage Materials   Materials Research Forum LLC
Materials Research Foundations **78** (2020) 50-90   https://doi.org/10.21741/9781644900871-3

dehydrocyclization of alkanes and alkenes meanwhile increasing the rate of the aromatization process. Zn-L was effective in catalyzing the Diels–Alder reaction (cyclization), which was another important source of aromatics. Besides, some strong acid sites formed on the catalyst surface were effective in promoting the hydrogen transfer reaction, which yielded surface functionalities. Hence, $ZnCl_2$ treatment resulted in the formation of abundant acid sites (mainly Lewis acid sites), whose catalytic effect led to the increase of the aromatic content and formation of surface functionalities in the activation. With increasing $ZnCl_2$ impregnation ratio, the new pores may be formed largely with the widened existing pores [76].

However, excessive $ZnCl_2$ addition may destroy significantly the micropore structure but enhance the mesoporous formation, thus leading to decline in surface area. On the other hand, $ZnCl_2$ can also restrict the tar formation. When the reaction temperature is < 500 °C, more volatiles are released by increasing the reaction temperature, and the surface area increases. However, at reaction temperature > 500 °C, this decrease in surface area of carbon can be attributed to the heat shrinking [77]. Possible catalytic reaction mechanism is summarized in Fig. 8.

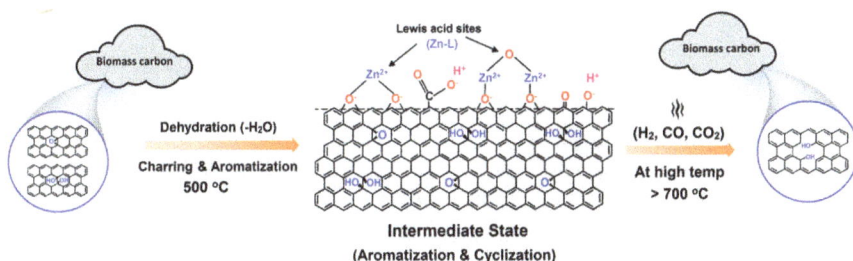

Figure 8. Possible reaction mechanism of ZnCl2 activation.

ACs prepared from renewable resources and biomass feedstocks, such as water bamboo [78], yeast cells [79], fallen leaves [80], loofah sponge [81], willow catkins [82], celtuce leaves [83], waste tea-leaves [84], cardamom pods [85], ginkgo shells [86], cow dung [87], silk [88], human hair [89], and sewage sludge [90], have been successfully produced for application of supercapacitors. For example, a highly porous carbon has been generated from cardamom pods (*Elettaria cardamomum L*) via $ZnCl_2$ activation at high temperature, showing a large SSA of 1298 m^2 g^{-1} and a pore volume of 1.23 cm^3 g^{-1} [85]. Organometallic precursor $[Re_2(CO)_{10}]$ may be reduced to Re NPs upon MW irradiation. The well-characterized Re@CDACs nanocomposites were then utilized for

Biomass Based Energy Storage Materials   Materials Research Forum LLC
Materials Research Foundations **78** (2020) 50-90   https://doi.org/10.21741/9781644900871-3

electrocatalytic oxidation of SY and supercapacitor applications (Fig. 9). The term "CDAC" denotes cardamom derived ACs. Moreover, these Re@CDACs catalysts yielded 181 F g^{-1} for C_{sp} at 1.6 A g^{-1} current density in 1.0 M H_2SO_4 electrolyte. The C_{sp} of 90% was retained after 2500 cycles at 2.0 A g^{-1}.

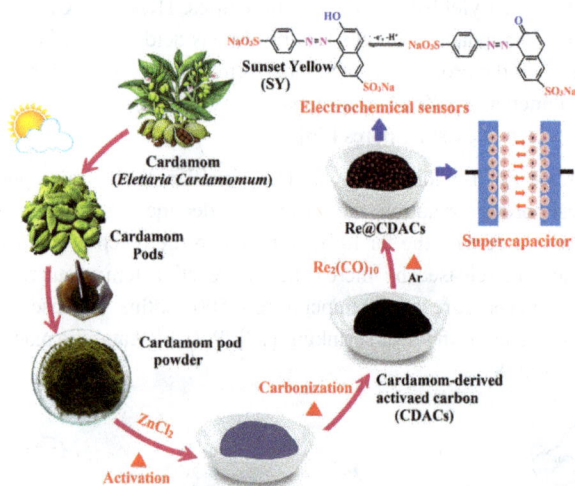

Figure 9 Schematic illustration for the preparation and applications of Re@CDAC composite [85]. Reproduced with permission from ref [85]. Copyright©2018 Elsevier publications.

Wang et al [83]. reported a celtuce leaf (CL)-derived porous carbon to have an ultrahigh SSA of 3404 m^2 g^{-1} and a large pore volume of 1.88 cm^3 g^{-1}. In the process of the KOH activation, the change of the chemical character and the porous structure of CL may significantly affect its electrochemical capacitive performance and CO_2 adsorption capacity at ambient pressure. CLs could act as new biomass source of porous carbon materials for high-performance supercapacitors and CO_2 capture (see Fig. 10a). The porous carbon-based supercapacitor electrode yielded large C_{sp} of 421 and 273 F g^{-1} in three and two-electrode systems, respectively. The KOH activation could generate nanoscale pores in the resultant carbon. Likewise, Hou et al [88]. reported another carbon-based nitrogen-doped hierarchical porous nanosheets (HPNC-NSs) through $FeCl_3$ and $ZnCl_2$ activation and then graphitization process. Accordingly, the HPNC-NS featured a large SSA of 2494 m^2 g^{-1} with hierarchical pores of 2.28 cm^3 g^{-1} to facilitate

Materials Research Forum LLC

https://doi.org/10.21741/9781644900871-3

rapid charge transfer, high charge capacity, adequate mass transport, easy electrolyte access, and minimized polarization effects (see Fig. 10b). Moreover, the silk derived hierarchical porous carbon in supercapacitor application reached a large C_{sp} of 242 F g^{-1}, energy density of 102 Wh kg^{-1}, and excellent cycling life stability [87].

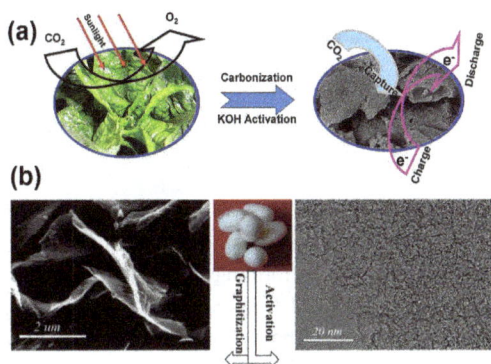

Figure 10. (a) The porous carbon derived from celtuce leaves for supercapacitor application. Adapted from Wang et al [83], and (b) Hierarchical porous nitrogen-doped carbon nanosheets prepared from natural silk via simultaneous activation and graphitization [88]. Reproduced with permission from ref [88]. Copyright©2015 The American Chemical Society.

Qian et al. [89] have synthesize micro/mesoporous carbon (HMC) materials of high charge storage capacity from human hair. The cleaned and dried hair fibers were cut into fine debris and pre-carbonized at 300 °C for 90 min. This pre-treated carbon material was then mixed with KOH (weight$_{KOH}$:weight$_{carbon}$ = 2) followed by carbonization at respective 700, 800, and 900 °C. The resultant carbon materials were characterized for the related morphology and chemical composition. Human hair carbonized at 800 °C yielded a C_{sp} of 340 F g^{-1} in 6 M KOH at a current density of 1 A g^{-1} and good stability over 20, 000 cycles. In contrast, a smaller value of 126 F g^{-1} at a current density of 1 A g^{-1} was obtained in a 1 M LiPF$_6$ ethylene carbonate/diethyl carbonate electrolyte. The factors of micro/mesoporosity in conjunction with large surface area and heteroatom doping effects, double layer and Faradaic contributions could play vital roles for enhancing supercapacitor behavior (see Fig. 11).

Biomass Based Energy Storage Materials
Materials Research Foundations **78** (2020) 50-90

Materials Research Forum LLC
https://doi.org/10.21741/9781644900871-3

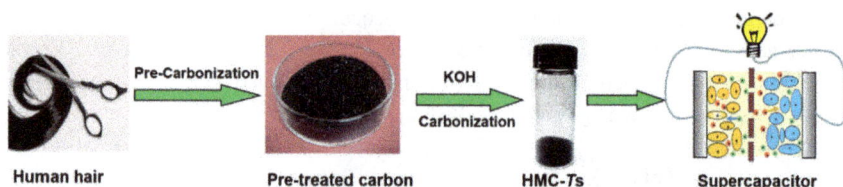

Figure 11. A flow diagram for the fabrication of HMCs [89]. Reproduced with permission from ref [89]. Copyright©2014 The Royal Society of Chemistry.

Recently, high-ash-content sewage sludge (SS) was used as a precursor to synthesize three-dimensional honeycomb-like hierarchically structured carbon (HSC) [90]. The fly-silicon process played a crucial role in forming honeycomb-like hierarchical structures. Fig. 12 illustrates the HSC synthesized from SS and activated sludge-derived carbons (ASC) without fly-silicon treatment. The yellowish-brown SS powder (Fig. 12a) was turned into a brownish-black powder (SC) with *finer* particles of about 12 mm average size after carbonization at high temperature (Fig. 12b). SC was found to contain the same quartz SiO_2 as in SS for its main crystalline phase. To remove the inorganic ash component (i.e. SiO_2), the resultant SC was treated by fly-silicon with hydrofluoric acid (HF) to obtain a black FSC powder (Fig. 12c). The FSC was then activated at 700 °C for 3 h with KOH at various KOH/C mass ratios to finally obtain 3D honeycomb-like HSC after HCl washing (Fig. 12d). This type of HSC exhibited excellent electrochemical performance containing surface area of 2839 m^2 g^{-1}, pore volume of 2.65 cm^3 g^{-1}, and capacitance of 379 F g^{-1} at 20 A g^{-1} in 6.0 M KOH electrolyte. 3D honeycomb-like HSC was demonstrated to maintain 90% of its initial capacitance after 20 000 cycles conducted at 20 A g^{-1}. Moreover, the assembled HSC//HSC symmetric supercapacitor further strengthens supercapacitive behavior with an energy density of 30.5 W h kg^{-1} in aqueous solution.

4.3   Hydrothermal carbonization

The hydrothermal carbonization (HTC) process yields a partially carbonized product so-called hydrochar with a low degree of condensation and a large amount of oxygen-containing groups [91]. The biomass-based nanostructured carbons may feature controlled morphology, appropriate functionality, and modified surface chemistry [92]. Low-temperature heating (180–250 °C) in sealed autoclave can enhance the structure aromaticity and carbon yield. Meanwhile, the SSA and surface chemistry of the carbon products could be further modified by addition of special oxidizing agent for chemical activation [93]. These modified SCs include cellulose, potato starch, eucalyptus wood sawdust [94], hemicellulose [95], paper pulp mill sludge [96], D-glucosamine [97], fungus [98], bamboo waste [99], microalgae [100], watermelon [101], and hemp [102].

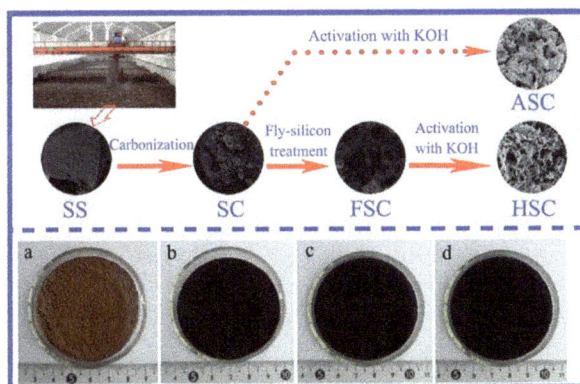

Figure 12. (Top) The synthesis process of 3D HSC with fly-silicon treatment and ASC without fly-silicon treatment. (Down) Photos of (a) sewage sludge, (b) sludge-derived carbon, (c) SC with fly-silicon treatment, and (d) the HSC materials by fly-silicon treatment and activation with KOH [90]. Reproduced with permission from ref [90]. Copyright©2015 The Royal Society of Chemistry.

Long et al. [98] reported a low cost approach to high quality of densely porous graphene-like carbon (PGC) materials which were synthesized through hydrothermal treatment of fungus (*Auricularia*). KOH was introduced to act as an in-built template that prevented the adjacent cell walls from fusion/agglomeration. Meanwhile, the KOH as an activating agent played a role to construct the porous carbon framework (see Fig 13), obtaining a high SSA (1103 m^2 g^{-1}) with high bulk density (about 0.96 g cm^{-3}) for the layer-stacking PGC. Such a porous framework of PGC offered more storage sites and short transport paths for electrolyte ions to enhance electrode conductivity. In this manner, the PGC electrode yielded a large volumetric capacitance of 360 F cm^3 and 99% capacitance retention after 10 000 cycles. Further, the assembled symmetric supercapacitor provided a volumetric energy density of 21 Wh L^{-1} with 96% C_{sp} retention after 10 000 cycles. A low cost and environmentally friendly electrode materials for superior volumetric capacitance performance become probable.

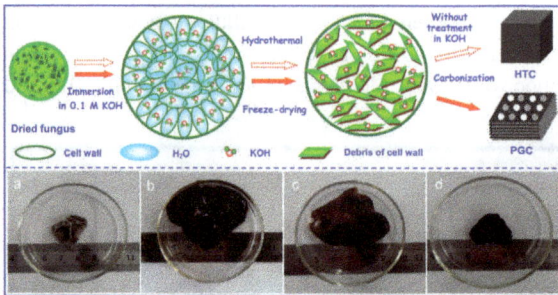

Figure 13. (Up) Schematic illustration of the HTC formation without KOH treatment and porous graphene-like carbon (PGC) materials. (Down): (a) The original dried fungus, (b) Swelled fungus after hydrothermal treatment in 0.1 M KOH solution, (c) Freeze-dried fungus and (d) The carbon materials by carbonization of freeze-dried fungus. Reproduced with permission from ref [98]. Copyright© 2015 Elsevier Publications.

For fabricating supercapacitor electrode material, Tian and co-workers [99] developed the beehive-like hierarchical nanoporous carbon (BHNC) obtaining SSA of 1472 m^2 g^{-1} and electronic conductivity of 4.5 S cm^{-1}. The HTC synthesis approach of BHNC is illustrated in Fig. 14a. The FESEM and TEM images show that BHNC has beehive-like hierarchical nanoporous structure, in which the micropores were found on the meso/macropores walls of the interconnected carbon nanosheet frameworks. The BHNC framework is shown with the structural model in Fig. 14a, and demonstrated in Fig. 14b–d and in Fig. 14e marked by the yellow rings. Fig.14f shows the partial graphitization from the distorted lattice fringe image of the BHNC. Moreover, as a remarkable supercapacitor electrode material, the BHNC sample yields a high C_{sp} of 301 F g^{-1} at 0.1 A g^{-1}, and a high P of 26 000 W kg^{-1} at an energy density of 6.1 W h kg^{-1}, but reduces to 192 F g^{-1} at 100 A g^{-1} with negligible capacitance loss after 20 000 cycles at 1 A g^{-1}.

Sevilla et al. [100] produced microporous carbons from the microalgae (as a N-rich carbon precursor) by the HTC process combined with a post-KOH activation treatment (Fig. 15). In this process, microalgae was converted to carbon material containing 0.7-2.7 wt.% of N content, a high SSA of ~2200 m^2 g^{-1} and high C_{sp} ~200 F g^{-1} with excellent rate capability.

As reported by Wu et al. [104], various porous carbons (PCs) were obtained from Shengli lignite (SL) via HTC treatment and KOH activation. The PCs had mainly micropores structure obtaining the highest SSA up to 3162 m^2 g^{-1}, and supercapacitor of 295 F g^{-1} at 40 mA g^{-1} which decreased to 210 F g^{-1} at 10 A g^{-1} in 6 M KOH electrolyte, retaining almost the same C_{sp} even after 15 000 cycles. Moreover, functional carbon materials with a large amount of oxygenated functional groups may be produced by the HTC synthesis method. The hydrochar rich in oxygenated functional groups makes it suitable for the applications such as adsorption, drug delivery, catalyst supports, and so on [105,106].

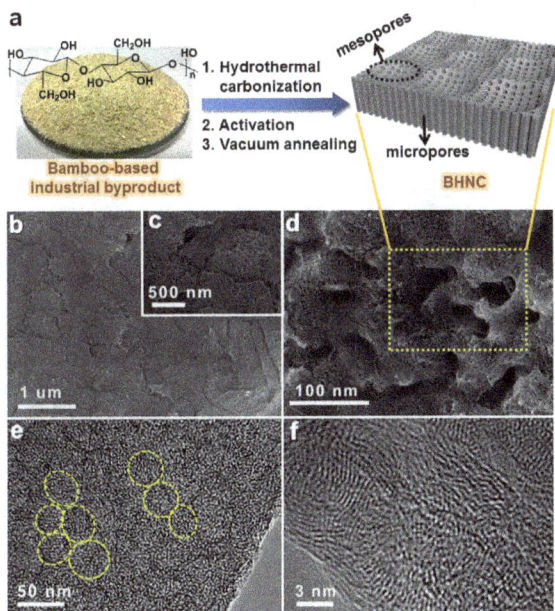

Figure 14. (a) Schematic of synthesis procedures of the BHNC sample, (b–d) FESEM, (e) TEM, and (f) HR-TEM image of the edge of the BHNC. Reproduced with permission from ref [99]. Copyright© 2015 The Royal Society of Chemistry.

4.4   Other activations

Activation is one of the most effective methods to enhance the surface area of carbon materials and regulate the meso/micropore proportion. During the past few years, most of PCs have been produced by aforementioned methods under an inert atmosphere [107]. Other activation methods inculding microwave carbonization (MC) [108,109], molten salt synthesis (MSS) [110], ionothermal carbonization (ITC) [111], and self-activation (SA) [112,113].

In MC activation approach the reactants are heated by MWs on the molecular level. The MC method has the merits of reduced operation time, improved energy efficiency and energy saving. For the conventional physical/chemical activation methods, the heat source was put outside the carbon bed, and thus a temperature gradient was generated from the hot outer to the interior, which easily caused distorted and inhomogeneous microstructure. In contrast, the temperature gradient with the MC treatment is reversed, making the MC method more effective to shorten processing time and to save higher energy [114].

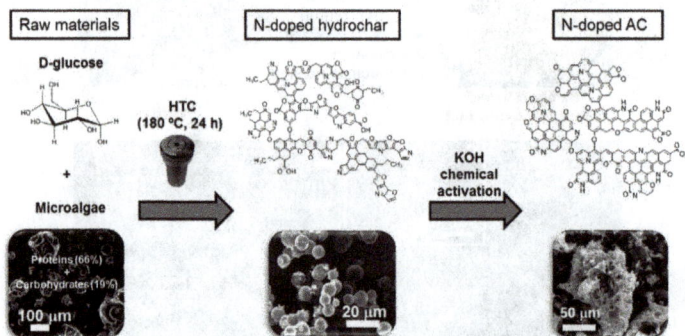

Figure 15. Synthesis of porous carbon from a mixture of microalgae and glucose by a coupled process of HTC and KOH activation. Reproduced with permission from ref [100]. Copyright© 2014 Elsevier publications.

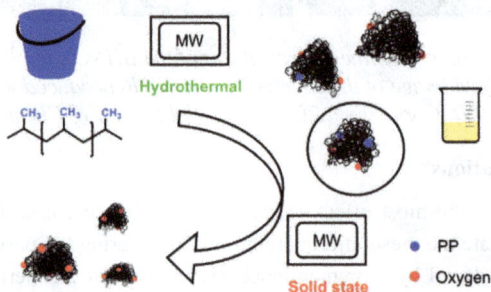

Figure 16. Schematic description of the MW-assisted hydrothermal treatment of PP and postmodification in air atmosphere by a solid state MW process [108]. Reproduced with permission from ref [108]. Copyright©2018 The American Chemical Society.

Recently, Adolfsson [108] demonstrated an upcycling of polypropylene (PP) waste to carbon materials by an MW assisted hydrothermal treatment (Fig 16). Similarly, Hu et al. [109]. successfully prepared CFs with good adsorptive capacity. However, the pre-oxidized fibers were carbonized at the temperature ranging from 400 to 1300 °C for 1 h. The CFs obtained at 710 °C with a thickness of 2 mm, exhibited the best MW absorbing ability. Hence it is important to obtain carbonaceous material from biomass by MC, which can be applied for supercapacitor electrode fabrications.

Jin-hui and co-workers [115] have compared the yield and porous nature of the AC prepared by the conventional heating and MW method. Jatropha hull was activated using the popular activating agents, such as steam and CO_2 activation. The scheme of the experimental apparatus using MW heating is shown in Fig 17.

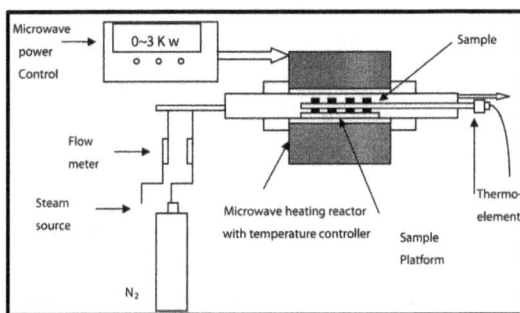

Figure 17. The diagram of apparatus for the preparation of activated carbons with MW [115]. Reproduced with permission from ref [115]. Copyright©2011 Elsevier publications.

Molten salt synthesis (MSS), as one of the biomass activation processes, implies genera of inorganic salts (metal halides or metal oxides) mixed with biomass to obtain the resultant 3D carbon materials [116]. The common salt agents employed particularly for low-melting-point salt (no more than 1000 °C) are such as zinc (potassium, sodium) chloride, carbonates, and nitrates. They create a liquid-flow phase for sufficient mass transfer and contact area at the solid–liquid interface [117]. Thus far, the complete mechanism is yet to know. The molten salts are believed to play a role in fabrication of biomass-derived carbon materials with appropriate pore size distributions [118]. For example, Wang and his co-workers converted cornstalk [119] into hierarchical porous carbon sheets (HPCS) and used as a pseudocapacitor electrode, which showed good performance. Fig. 18 shows the schematic diagram for the preparation of HPCS under air.

Biomass Based Energy Storage Materials   Materials Research Forum LLC
Materials Research Foundations **78** (2020) 50-90   https://doi.org/10.21741/9781644900871-3

Firstly, the precursor was covered with NaCl and KCl mixed salt during the activation process and the contents were calcined in a muffle furnace at 800 $^\circ$C for 3 h. After removing the common salt, such HPCS contain an ultra-thin sheet structure and abundant hierarchical pores to facilitate electrolyte diffusion and ion transfer [119].

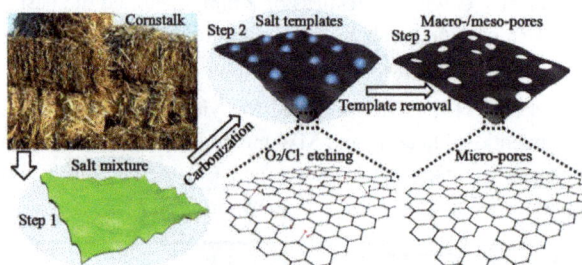

Figure 18. Schematic illustration of the synthesis process for HPCS from cornstalk under an air atmosphere [119]. Reproduced with permission from ref [119]. Copyright©2018 The Royal Society of Chemistry.

In addition, the HPCS exhibited large SSA of 1588 m^2 g^{-1} with a large value of 407 F g^{-1} at 1 A g^{-1} in a three-electrode system. The assembled symmetric supercapacitor in a two-electrode system also yielded 413 F g^{-1} at 0.5 A g^{-1}, with excellent rate capacity and 92.6% of initial capacitance retention after 20 000 cycles at 5 A g^{-1}. The salt agents as well as the ratio of the carbon precursor to the salt should be carefully selected because the end product of carbon and its properties depend largely on the different molten salt systems.

Low cost carbon materials with interconnected, multichannel, and porous structures could be carbonized at high temperatures under inert condition. Cheng et al. [110] employed rapid MC coupled with MSS method to successfully convert plant-derived biomass (i.e., carrot) to nitrogen and oxygen enriched hierarchically porous carbons (NOHPCs), as displayed in Fig. 19. The obtained sample possesses SSA of 1899 m^2 g^{-1} with a pore volume of 1.16 cm^3 g^{-1}, mesopore ratio ~70%, the heteroatom of nitrogen of 5.30 wt% and the oxygen content of 14.12 wt%. Herein, ZnCl$_2$ plays important roles as MW absorber, chemical activation agent, and porogen (pore generating solvent) in the process. The capacitive properties of the NOHPCs based electrodes were measured in 6 M KOH to exhibit 276 F g^{-1} at 0.2 A g^{-1}, with 90.9% C_{sp} retention at 10 A g^{-1}. Further, the symmetric supercapacitor based on NOHPC in 1 M Na$_2$SO$_4$ electrolyte exhibits a high energy density of 13.9 Wh kg^{-1} at a P of 120 W kg^{-1} and 95% capacitance retention after 10 000 cycles.

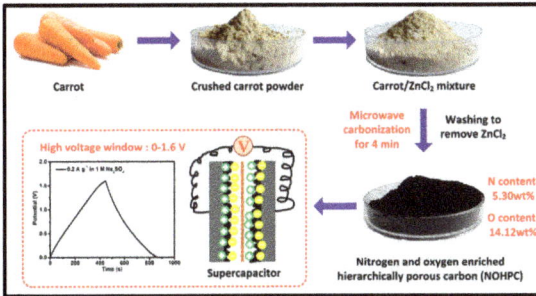

Figure 19. Molten salt synthesis NOHPCs derived from carrot via rapid MC for supercapacitors [110]. Reproduced with permission from ref [110]. Copyright©2018 Elsevier publications.

Figure 20. Preparation procedure of hierarchical porous carbons [121]. Reproduced with permission from ref [121]. Copyright©2017 The Royal Society of Chemistry.

Ionothermal carbonization (ITC) technique is another activation approach to synthesize hetero atoms rich carbons, enabling biomass conversion to functional carbonaceous materials under relatively mild conditions. As such, abundant heteroatoms can be doped into the final carbon framework by selecting specific biomass as precursor [120]. Recent work has suggested that ITC is one of the most efficient methods for PC preparation. For example, Liu et al. [121] have reported use of nitrogen-containing Jujun grass as the carbon precursor, because of its wide availability, low cost and ability to introduce

71

nitrogen atoms into final products. The synthesis procedure is illustrated in Fig 20. In addition, it possesses high SSA (S_{BET} = 2532 m^2 g^{-1}) and abundant mesopores (V_{meso} = 1.077 cm^3 g^{-1}). On the other hand, the ionic liquid (1-butyl-3-methylimidazolium tetrachloroferrate [Bmim][FeCl$_4$]) acted as a reaction medium for the carbon conversion as a porogenic agent for inducing mesoporosity. Moreover, the assembled supercapacitor also yielded 336 F g^{-1} at 1 A g^{-1} in 6 M KOH and retained 222 F g^{-1} even at 10 A g^{-1}. The maximum energy density of PCs was found to be over 72.7 W h kg^{-1} when the P was 1204 W kg^{-1}, higher than most equivalent benchmarks in aqueous electrolytes.

Self-activating the biomass does not need additional activation reagents to decrease the cost and simplify the procedure [112–114]. However, the biomass needs high content of potassium or sodium-based inorganic salts and the products often suffer from low SSA, which limits its application [122]. Also, this process uses gases emitted from the biomass pyrolysis to activate the converted carbon (see Fig. 21).

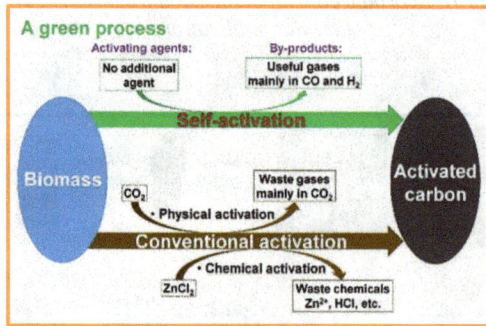

Figure 21. Self-activation for activated carbon from biomass. Reproduced with permission from ref [122]. Copyright©2016 The Royal Society of Chemistry.

So far, the biomass-based PCs materials adopted are mostly toxic and corrosive activation agents along multi-step processes, resulted in low yield as well as low SSA. Therefore, a great challenge is encountered for achieving a high yield of porous carbon with a large surface area, without sacrificing the natural structure and immanent heteroatoms (e.g., N and S) of the biomass. The presence of heteroatoms such as N and S enhanced the electrochemical properties of PCs. Thus, looking for renewable carbon sources along with developing simple and easy synthesis routes is essential and highly needed [123].

Shi et al. [124] designed self-activation fabrication method to address these current issues of conventional activations. Self-activation, as schematically illustrated in Fig. 22a, utilizes the emitted gases during carbonization process to serve as activating agents [125]. When pine wood was pyrolyzed at 1050 ˚C, various emitted gases (CO, H_2, CO_2, CH_4) evolved during the activation process of the carbonized wood, obtaining SSA of 2738 m^2 g^{-1} and specific pore volume (V_{sp}) of 2.209 cm^3 g^{-1}. The self-activation of kenaf core [122] was found that the pore expansion dominated in the early pyrolysis stage, and then the pore combination increased as the pyrolysis continued (Fig. 22b). Eventually, the residual AC would turn into ash because of the pore combination.

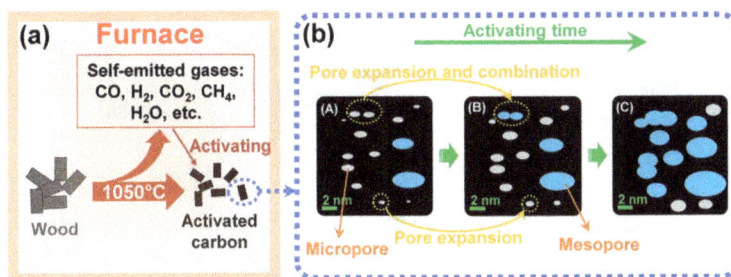

Figure 22. Illusions of (a) self-activation process, and (b) changes of pores with increasing activating time. Reproduced with permission from ref [125]. Copyright©2016 Wiley-VCH Verlag GmbH & Co. KGaA, Weinheim.

Self-activation has great potential to be developed into next-generation of high-performance electrode materials for electrochemical energy storage devices [125,126]. Some excellent reviews highlighting the performance of various biomass-derived carbon materials have been appeared [127–130]. We compiled the reports on some biomass-derived carbons employed for supercapacitor applications is presented in Table 1.

Table 1. Comparison of biomass-derived activated carbon for supercapacitor applications

Carbon Source	Photograph	Electrolyte/ concerntration	Potential window (V)	Stability	Current density (A g^{-1})	C_{sp} (F g^{-1})	E (Wh kg^{-1})	Ref
Sugarcane bagasse		H$_2$SO$_4$	0.0 to 1.0	5,000	0.25	300	10	[44]
Camellia oleifera		KOH/6M	-0.2 to 0.8	5,000	0.2	374	–	[52]
Paper flower		H$_2$SO$_4$/1M	*-0.1 to 0.9*	10,000	1.0	118	–	[71]
Bamboo		KOH/6M	0.0 to 1.0	20,000	0.1	301	6	[99]
Watermelon		KOH/6M	-1.0 to 0.0	10,000	0.5	348	11.3	[131]
Grape fruit peel		H$_2$SO$_4$/1M	0.0 to 1.1	10,000	0.1	311	34	[132]
Banana fiber		KOH/	–	5,000	10 mV s^{-1}	324	63	[133]
Coffee beans		H$_2$SO$_4$/1M	0.0 to 1.0	10,000	0.05	368	20	[134]
Banana peel		KOH/6M	-1.0 to 0.0	–	1.0	206	–	[135]
Paulownia flower		KOH/6M	-1.0 to 1.0	–	1.0	297	22.2	[136]
Carrageenan sea weed		KOH/6M	-0.1 to -1.1	20,000	1.0	230	–	[137]
Lignin		KOH/6M	0.0 to 1.0	20,000	1.0	312	44.7	[138]
Corncob		H$_2$SO$_4$/1M	-0.8 to 0.4	10,000	0.5	390	25	[139]
Coconut shell and sewage sludge		KOH/6M	-1.0 to 1.0	10,000	0.5	420	25.4	[140]

5.   Outlook

This book chapter highlights the important developments of porous carbon (PC) nanomaterials for their most distinguished SC applications. PC have been widely used in

Biomass Based Energy Storage Materials   Materials Research Forum LLC
Materials Research Foundations **78** (2020) 50-90   https://doi.org/10.21741/9781644900871-3

sensors, fuel cell, adsorption, solar cell, catalysis, and energy-related applications. In biomass conversion to carbon nanomaterials, much progress has been achieved in the synthesis and characterization of these PC nanomaterials and their applications to various energy fields. Currently, the PC materials as the main choice for SC electrodes are in high demand, owing of their large SSA and pore volumes, active sites, low toxicity, and chemically modifiable surfaces. The application of activation methods has universal significance in producing highly porous carbons for various high-performance energy storage/conversion applications. The preparation of functional PC nanomaterials-related chemistry helps to design an electrode material in energy storage devices.

Various surface functional groups are generated upon the bonding or doping of heteroatoms such as N, O, P, S, B, etc., into carbon scaffold to improve the energy storage, because these heteroatoms allow the rapid occurrence of redox processes. Although new routes may open toward the rational optimization of efficient catalysts, more predictive capacity is required with theoretical methods to fulfill the concept of sustainable chemistry. Moreover, their performance such as energy density and retention rate still needs to be further improved by porosity regulation and surface modification through novel activation methods. We believe that activation of biomass and fabrication of carbon electrode based SC will have a promising future.

Conclusions and prospects

Biomass is suitable for producing functional porous carbon materials in application of electrocapacitive energy storage. As comparable to commercial AC, biomass-derived carbon materials are prepared by simple carbonization and/or activation methods showing excellent electrocapacitive performance. Advanced hierarchical porous carbons can be obtained by using the chemical activation methods such as KOH. Challenges and issues have been addressed to enable carbon materials thus obtained to find practical applications in energy storage. This chapter provides a holistic overview of electrochemical energy storage devices using plant-based biomass from a cross-disciplinary perspective that encompasses materials science, chemical engineering, and environmental engineering.

Abbreviations and Acronyms

PCs   Porous carbons
ACs   Activated carbons
OMCs   Ordered mesoporous carbon
CFs   Carbon fibers
CNTs   Carbon nanotubes

EDLC	Electrical double layer capacitance
SSC	Symmetric supercapacitor
ASC	Asymmetric supercapacitor
SSA	Specific surface areas
C_{sp}	Specific capacitance
E	Energy density
P	Power density
HPCS	Hierarchical porous carbon sheets
MC	Microwave carbonization
MSS	Molten salt synthesis
SA	Self-activation
ITC	Ionothermal carbonization

References

[1] M.-M. Titirici, R.J. White, N. Brun, V.L. Budarin, D.S. Su, F. Monte, J.H. Clark, M.J. MacLachlan, Sustainable carbon materials, Chem. Soc. Rev. 44 (2015) 250–290. https://doi.org/10.1039/C4CS00232F

[2] M.R. Benzigar, S. N. Talapaneni, S. Joseph, K. Ramadass, G. Singh, J. Scaranto, U. Ravon, K. Al-Bahily, A. Vinu, Recent advances in functionalized micro and mesoporous carbon materials: synthesis and applications, Chem. Soc. Rev. 47 (2018) 2680–2721. https://doi.org/10.1039/C7CS00787F

[3] L. Estevez, D. Barpaga, J. Zheng, S. Sabale, R.L Patel, J.G. Zhang, B.P. McGrail, R.K. Motkuri, Hierarchically porous carbon materials for CO_2 capture: The role of pore structure, Ind. Eng. Chem. Res. 57 (2018) 1262–1268. https://doi.org/10.1021/acs.iecr.7b03879

[4] A. Sahasrabudhe, S. Kapri, S. Bhattacharyya, Graphitic porous carbon derived from human hair as 'green' counter electrode in quantum dot sensitized solar cells, Carbon 107 (2016) 395–404. https://doi.org/10.1016/j.carbon.2016.06.015

[5] Beguin, F. (Ed.), Frackowiak, E. (Ed.). Carbons for electrochemical energy storage and conversion systems, Boca Raton: CRC Press, 2010. https://doi.org/10.1201/9781420055405

[6] A. Yu, V. Chabot, J. Zhang, Electrochemical supercapacitors for energy storage and delivery fundamentals and applications, CRC Press, 2013.

[7] Y. Wang, Y. Song, Y. Xia, Electrochemical capacitors: mechanism, materials, systems, characterization and applications, Chem. Soc. Rev. 45 (2016) 5925–5950. https://doi.org/10.1039/C5CS00580A

[8]   A. Burke, R&D considerations for the performance and application of electrochemical capacitors, Electrochim. Acta 53 (2007) 1083–1091. https://doi.org/10.1016/j.electacta.2007.01.011

[9]   T. Brousse, M. Toupin, D. Belanger,A hybrid activated carbon-manganese dioxide capacitor using a mild aqueous electrolyte, J. Electrochem. Soc. 152 (2004) A614–A622. https://doi.org/10.1149/1.1650835

[10]   F.X. Ma, L. Yu, C.Y. Xu, X.W. Lou, Self-supported formation of hierarchical $NiCo_2O_4$ tetragonal microtubes with enhanced electrochemical properties, Energy Environ. Sci. 9 (2016) 862–866. https://doi.org/10.1039/C5EE03772G

[11]   X.Y. Yu, L. Yu, X.W. Lou, Metal sulfide hollow nanostructures for electrochemical energy storage, Adv. Energy Mater. 6 (2016) 1501333. https://doi.org/10.1002/aenm.201501333

[12]   L. Yu, B. Guan, W. Xiao, X.W. Lou, Formation of yolk-shelled Ni–Co mixed oxide nanoprisms with enhanced electrochemical performance for hybrid supercapacitors and lithium ion batteries, Adv. Energy Mater. 5 (2015) 1500981. https://doi.org/10.1002/aenm.201500981

[13]   Y. Guo, L. Yu, C.Y. Wang, Z. Lin, X.W. Lou, Hierarchical tubular structures composed of Mn-based mixed metal oxide nanoflakes with enhanced electrochemical properties, Adv. Funct. Mater. 25 (2015) 5184–5189. https://doi.org/10.1002/adfm.201501974

[14]   Y.M. Chen, Z. Li, X.W. Lou, General formation of $M_xCo_{3-x}S_4$ (M=Ni, Mn, Zn) hollow tubular structures for hybrid supercapacitors, Angew. Chem. Int. Ed. 54 (2015) 10521–10524. https://doi.org/10.1002/anie.201504349

[15]   L. Liu, Z. Niu, J. Chen, Flexible supercapacitors based on carbon nanotubes, Chin. Chem. Lett. 29 (2018) 571–581. https://doi.org/10.1016/j.cclet.2018.01.013

[16]   A.M. Abioye, F.N. Ani, Recent development in the production of activated carbon electrodes from agricultural waste biomass for supercapacitors: A review, Renew. Sust. Energ. Rev. 52 (2015) 1282–1293. https://doi.org/10.1016/j.rser.2015.07.129

[17]   P. Kossyrev, Carbon black supercapacitors employing thin electrodes, J. Power Sources 201 (2012) 347–352. https://doi.org/10.1016/j.jpowsour.2011.10.106

[18]   X. Yu, J.G. Wang, Z.H. Huang, W. Shen, F. Kang, Ordered mesoporous carbon nanospheres as electrode materials for high-performance supercapacitors, Electrochem. Commun. 36 (2013) 66–70. https://doi.org/10.1016/j.elecom.2013.09.010

[19]   P. Suktha, P. Chiochan, P. Iamprasertkun, J. Wutthiprom, N. Phattharasupakun, M. Suksomboon, T. Kaewsongpol, P. Sirisinudomkit, T. Pettong, M. Sawangphruk, High-performance supercapacitor of functionalized carbon fiber paper with high surface ionic and bulk electronic conductivity: Effect of organic functional groups, Electrochim. Acta 176 (2015) 504–513. https://doi.org/10.1016/j.electacta.2015.07.044

[20]   F.Y. Zeng, Z.Y. Sui, S. Liu, H.P. Liang, H.H. Zhan, B.-H. Han, Nitrogen-doped carbon aerogels with high surface area for supercapacitors and gas adsorption, Mater. Today Commun. 16 (2018) 1–7. https://doi.org/10.1016/j.mtcomm.2018.03.015

[21]   H. Sheng, M. Wei, A. D'Aloia, G. Wu, Heteroatom polymer-derived 3D high-surface-Area and mesoporous graphene sheet-like carbon for supercapacitors, ACS Appl. Mater. Interfaces 8 (2016) 30212–30224. https://doi.org/10.1021/acsami.6b10099

[22]   Q. Ke, J. Wang, Graphene-based materials for supercapacitor electrodes A review, J. Materiomics 2 (2016) 37–54. https://doi.org/10.1016/j.jmat.2016.01.001

[23]   S. Yue-feng, W. Feng, B. Liying, Y. Zhao-hui, RuO_2/activated carbon composites as a positive electrode in an alkaline electrochemical capacitor, New Carbon Mater. 22 (2007) 53–58. https://doi.org/10.1016/S1872-5805(07)60007-9

[24]   P. Jeżowski, K. Fic, O. Crosnier, T. Brousse, F. Béguin, Lithium rhenium(VII) oxide as a novel material for graphite pre-lithiation in high performance lithium-ion capacitors, J. Mater. Chem. A 4 (2016) 12609–12615. https://doi.org/10.1039/C6TA03810G

[25]   M.S. Yadav, S.K. Tripathi, Synthesis and characterization of nanocomposite NiO/activated charcoal electrodes for supercapacitor application, Ionics 23 (2017) 2919–2930. https://doi.org/10.1007/s11581-017-2026-9

[26]   Q. Meng, K. Cai, Y. Chen, L. Chen, Research progress on conducting polymer based supercapacitor electrode, materials, Nano Energy 36 (2017) 268–285. https://doi.org/10.1016/j.nanoen.2017.04.040

[27]   C. Wang, T. Liu, Nori-based N, O, S, Cl co-doped carbon materials by chemical activation of $ZnCl_2$ for supercapacitor, J. Alloys. Compd. 696 (2017) 42–50. https://doi.org/10.1016/j.jallcom.2016.11.206

[28]   M.-S. Balogun, Y. Huang, W. Qiu, H. Yang, H. Ji, Y. Tong, Updates on the development of nanostructured transition metal nitrides for electrochemical energy storage and water splitting, Mater. Today 20 (2017) 425–451. https://doi.org/10.1016/j.mattod.2017.03.019

[29] N. Nitta, F. Wu, J. T. Lee, G. Yushin, Li-ion battery materials: present and future, Mater. Today 18 (2015) 252–264. https://doi.org/10.1016/j.mattod.2014.10.040

[30] H.C. Chang, H.Y. Chang, W.J. Su, K.Y. Lee, W.C. Shih, Preparation and electrochemical characterization of NiO nanostructure-carbon nanowall composites grown on carbon cloth, Appl. Surf. Sci. 258 (2012) 8599–8602. https://doi.org/10.1016/j.apsusc.2012.05.057

[31] Y. Xia, W. Zhang, Z. Xiao, H. Huang, H. Zeng, X. Chen, F. Chen, Y. Gan, X. Tao, Biotemplated fabrication of hierarchically porous NiO/C composite from lotus pollen grains for lithium-ion batteries, J. Mater. Chem. 22 (2012) 9209–9215. https://doi.org/10.1039/c2jm16935e

[32] R. Madhu, V. Veeramani, S.M. Chen, P. Veerakumar, S.-B. Liu, Functional porous carbon/nickel oxide nanocomposites as binder-free electrodes for supercapacitors, Chem. Eur. J. 21 (2015) 8200–8206. https://doi.org/10.1002/chem.201500247

[33] S.T. Senthilkumar, R. Kalai Selvan, J.S. Melo, The biomass derived activated carbon for supercapacitor, AIP Conf. Proc. 1538 (2013) 124–127. https://doi.org/10.1063/1.4810042

[34] Z. Gao, Y. Zhang, N. Song, X. Li, Biomass-derived renewable carbon materials for electrochemical energy storage, Mater. Res. Lett. 5 (2017) 69–88. https://doi.org/10.1080/21663831.2016.1250834

[35] Y.P. Gao, Z.-B. Zhai, K.J. Huang, Y.Y. Zhang, Energy storage applications of biomass-derived carbon materials: Batteries and supercapacitors, New J. Chem. 41 (2017) 11456–11470. https://doi.org/10.1039/C7NJ02580G

[36] R.B. Marichi, V. Sahu, R.K. Sharma, G. Singh, Efficient, sustainable, and clean energy storage in supercapacitors using biomass derived carbon materials, Springer International Publishing AG 2018 L.M.T. Martínez et al. (eds.), Handbook of Ecomaterials. https://doi.org/10.1007/978-3-319-48281-1_155-1

[37] J. Deng, M. Li, Y. Wang, Biomass-derived carbon: synthesis and applications in energy storage and conversion, Green Chem. 18 (2016) 4824–4854. https://doi.org/10.1039/C6GC01172A

[38] S. Zhou, L. Zhou, Y. Zhang, J. Sun, J. Wen, Y. Yuan. Upgrading earth-abundant biomass into three dimensional carbon materials for energy and environmental applications, J. Mater. Chem. A 7 (2019) 4217–4229. https://doi.org/10.1039/C8TA12159A

[39] A.M. Jacob, V.B. Igor, The potential of biomass in the production of clean transportation fuels and base chemicals. Production and purification of ultraclean transportation fuels: American Chem. Soc. (2011) 65–77. https://doi.org/10.1021/bk-2011-1088.ch005

[40] P. González-García, T.A. Centeno, E. Urones-Garrote, D. Ávila-Brande, L.C. Otero-Díaz, Microstructure and surface properties of lignocellulosic-based activated carbons. Appl. Surf. Sci. 265 (2013) 731–737. https://doi.org/10.1016/j.apsusc.2012.11.092

[41] R. Farma, M. Deraman, A. Awitdrus, I.A. Talib, E. Taer, N.H. Basri, J.G. Manjunatha, M.M. Ishak, B.N.M. Dollah, S.A. Hashmiet, Preparation of highly porous binder less activated carbon electrodes from fibres of oil palm empty fruit bunches for application in supercapacitors, Bioresour. Technol. 132 (2013) 254–261. https://doi.org/10.1016/j.biortech.2013.01.044

[42] A.E. Ismanto, S. Wang, F.E. Soetaredjo, S. Ismadji, Preparation of capacitor's electrode from cassava peel waste, Bioresour. Technol. 101 (2010) 3534–3540. https://doi.org/10.1016/j.biortech.2009.12.123

[43] B. Xu, Y. Chen, G. Wei, G. Cao, H. Zhang, Y, Yang, Activated carbon with high capacitance prepared by NaOH activation for supercapacitors, Mater. Chem. Phys. 124 (2010) 504–509. https://doi.org/10.1016/j.matchemphys.2010.07.002

[44] T.E. Rufford, D. Hulicova-Jurcakova, K. Khosla, Z. Zhu, G.Q. Lu, Microstructure and electrochemical double-layer capacitance of carbon electrodes prepared by zinc chloride activation of sugar cane bagasse, J. Power Sources 195 (2010) 912–918. https://doi.org/10.1016/j.jpowsour.2009.08.048

[45] W.J. Si, X.Z. Wu, W. Xing, J. Zhou, S.P. Zhuo, Bagasse-based nanoporous carbon for supercapacitor application, J. Inorg. Mater. 26 (2011) 107–112. https://doi.org/10.3724/SP.J.1077.2010.10376

[46] K.Y. Foo, B.H. Hameed, Utilization of rice husks as a feedstock for preparation of activated carbon by microwave induced KOH and K_2CO_3 activation. Bioresour. Technol. 102 (2011) 9814–9817. https://doi.org/10.1016/j.biortech.2011.07.102

[47] X. He, P. Ling, J. Qiu, M. Yu, X. Zhang, C. Yu, M. Zheng, Efficient preparation of biomass based mesoporous carbons for supercapacitors with both high energy density and high power density, J. Power Sources 240 (2013) 109–113. https://doi.org/10.1016/j.jpowsour.2013.03.174

[48] X. Li, W. Xing, S. Zhuo, J. Zhou, F. Li, S.Z. Qiao, G.Q Lu, Preparation of capacitor's electrode from sunflower seed shell, Bioresour. Technol. 102 (2011) 1118–1123. https://doi.org/10.1016/j.biortech.2010.08.110

[49] J.M. Valente Nabais, J.G. Teixeira, I. Almeida, Development of easy made low cost bindless monolithic electrodes from biomass with controlled properties to be used as electrochemical capacitors, Bioresour. Technol. 102 (2011) 2781–2787. https://doi.org/10.1016/j.biortech.2010.11.083

[50] E. Taer, M. Deraman, I.A. Talib, A. Awitdrus, S.A. Hashmi, A.A. Umar, Preparation of a highly porous binderless activated carbon monolith from rubber wood sawdust by a multi-step activation process for application in supercapacitors, Int. J. Electrochem. Sci. 6 (2011) 3301–3315.

[51] K.Y. Foo, B.H. Hameed, Preparation of oil palm (Elaeis) empty fruit bunch activated carbon by microwave-assisted KOH activation for the adsorption of methylene blue. Desalination 275 (2011) 302–305. https://doi.org/10.1016/j.desal.2011.03.024

[52] K.Y. Foo, B.H. Hameed, Utilization of oil palm biodiesel solid residue as renewable sources for preparation of granular activated carbon by microwave induced KOH activation, Bioresour. Technol. 130 (2013) 696–702. https://doi.org/10.1016/j.biortech.2012.11.146

[53] S. Bhoyate, C.K. Ranaweera, C. Zhang, T. Morey, M. Hyatt, P.K. Kahol, M. Ghimire, S.R. Mishra, R.K. Gupta, Eco-friendly and high performance supercapacitors for elevated temperature applications using recycled tea leaves. Global Challenges 1 (2017) 1700063. https://doi.org/10.1002/gch2.201700063

[54] H. Wang, H. Yi, X. Chen, X. Wang, Asymmetric supercapacitors based on nanoarchitectured nickel oxide/graphene foam and hierarchical porous nitrogen-doped carbon nanotubes with ultrahigh-rate performance, J. Mater. Chem. A 2 (2014) 3223–3230. https://doi.org/10.1039/C3TA15046A

[55] H. Lu, X.S. Zhao, Biomass-derived carbon electrode materials for supercapacitors, Sustain. Energ. Fuels 1 (2017) 1265–1281. https://doi.org/10.1039/C7SE00099E

[56] H. Yang, M. Yoshio, K. Isono, R. Kuramoto, Improvement of commercial activated carbon and its application in electric double layer capacitors, Electrochem. Solid-State Lett. 5 (2002) A141–A144. https://doi.org/10.1149/1.1477297

[57] H. Teng, Y-J.C. Chien, To hsieh performance of electric double-layer capacitors using carbons prepared from phenol-formaldehyde resins by KOH etching, Carbon 39 (2001) 1981–1987. https://doi.org/10.1016/S0008-6223(01)00027-6

[58] B. Xu, Y. Chen, G. Wei, G. Cao, H. Zhang, Y. Yang, Activated carbon with high capacitance prepared by NaOH activation for supercapacitors. Mater. Chem. Phys. 124 (2010) 504–509. https://doi.org/10.1016/j.matchemphys.2010.07.002

Biomass Based Energy Storage Materials
Materials Research Foundations **78** (2020) 50-90

Materials Research Forum LLC
https://doi.org/10.21741/9781644900871-3

[59] I.I. Gurten, M. Ozmak, E. Yagmur, Z. Aktas, Preparation and characterisation of activated carbon from waste tea using K_2CO_3, Biomass Bioenergy 37 (2012) 73–81. https://doi.org/10.1016/j.biombioe.2011.12.030

[60] S. Aber, A. Khataee, M. Sheydaei, Optimization of activated carbon fiber preparation from Kenaf using K_2HPO_4 as chemical activator for adsorption of phenolic compounds. Bioresour. Technol. 100 (2009) 6586–6591. https://doi.org/10.1016/j.biortech.2009.07.074

[61] M. Benadjemia, L. Millière, L. Reinert, N. Benderdouche, L. Duclaux, Preparation, characterization and methylene blue adsorption of phosphoric acid activated carbons from globe artichoke leaves. Fuel Process Technol. 92 (2011) 1203–1212. https://doi.org/10.1016/j.fuproc.2011.01.014

[62] B.S. Lou, P. Veerakumar, S.M. Chen, V. Veeramani, R. Madhu, S.-B. Liu, Ruthenium nanoparticles decorated curl-like porous carbons for high performance supercapacitors, Sci. Rep. 6 (2016) 19949. https://doi.org/10.1038/srep19949

[63] S. Karagoz, T. Tay, S. Ucar, M. Erdem. Activated carbons from waste biomass by sulfuric acid activation and their use on methylene blue adsorption, Bioresource Technol. 99 (2008) 6214–6222. https://doi.org/10.1016/j.biortech.2007.12.019

[64] P. Alvarez, C. Blanco, M. Granda, The adsorption of chromium (VI) from industrial wastewater by acid and base-activated lignocellulosic residues, J. Hazard. Mater 144 (2007) 400–405. https://doi.org/10.1016/j.jhazmat.2006.10.052

[65] K. Gadkaree, M. Jaroniec, Pore structure development in activated carbon honeycombs, Carbon 38 (2000) 983–993. https://doi.org/10.1016/S0008-6223(99)00204-3

[66] C. Moreno-Castilla, M. Ferro-Garcia, J. Joly, I. Bautista-Toledo, F. Carrasco-Marin, J. Rivera-Utrilla, Activated carbon surface modifications by nitric acid, hydrogen peroxide, and ammonium peroxydisulfate treatments, Langmuir 11 (1995) 4386–4392. https://doi.org/10.1021/la00011a035

[67] J.M. Valente Nabais, J.G. Teixeira, I. Almeida, Development of easy made low cost bindless monolithic electrodes from biomass with controlled properties to be used as electrochemical capacitors, Bioresour Technol. 102 (2011) 2781–2787. https://doi.org/10.1016/j.biortech.2010.11.083

[68] P. Veerakumar, T. Maiyalagan, B.G. Sundara Raj, K. Guruprasad, Z. Jiang K.-C, Lin, Paper flower-derived porous carbons with high-capacitance by chemical and physical activation for sustainable applications, Arab. J. Chem. 2018.

[69] F.C. Wu, R.L. Tseng, C.C. Hu, C.C. Wang, Effects of pore structure and electrolyte on the capacitive characteristics of steam and KOH-activated carbons for supercapacitors, J Power Sources 144 (2005) 302–309. https://doi.org/10.1016/j.jpowsour.2004.12.020

[70] P. Zhang, F. Sun, Z. Shen, D. Cao, ZIF-derived porous carbon: A promising supercapacitor electrode material, J. Mater. Chem. A 2 (2014)12873–12880. https://doi.org/10.1039/C4TA00475B

[71] A. Halama, B. Szubzda, G. Pasciak, Carbon aerogels as electrode material for electrical double layer supercapacitors synthesis and properties, Electrochim. Acta 55 (2010) 7501–7505. https://doi.org/10.1016/j.electacta.2010.03.040

[72] C. Wang, T. Liu, Nori-based N, O, S, Cl co-doped carbon materials by chemical activation of $ZnCl_2$ for supercapacitor, J. Alloy. Compd. 696 (2017) 42–50. https://doi.org/10.1016/j.jallcom.2016.11.206

[73] W. Chen, H. Zhang, Y. Huang, W. Wang, A fish scale based hierarchical lamellar porous carbon material obtained using a natural template for high performance electrochemical capacitors, J Mater Chem. 20 (2010) 4773–4775. https://doi.org/10.1039/c0jm00382d

[74] D. Kalpana, S.H. Cho, S.B. Lee, Y.S. Lee, R. Misra, N.G. Renganathan, Recycled waste paper–a new source of raw material for electric double-layer capacitors, J. Power Sources 190 (2009) 587–591. https://doi.org/10.1016/j.jpowsour.2009.01.058

[75] K. Sun, Q. Huang, Y. Chi, J. Yan,Effect of $ZnCl_2$-activated biochar on catalytic pyrolysis of mixed wasteplastics for producing aromatic-enriched oil, Waste Manage. 81 (2018) 128–137. https://doi.org/10.1016/j.wasman.2018.09.054

[76] R. Chen, L. Li, Z. Liu, M. Lu, C. Wang, H. Li,W. Ma, S. Wang, Preparation and characterization of activated carbons fromtobacco stem by chemical activation, J. Air. Waste. Manag. Assoc. 67 (2017)713–724. https://doi.org/10.1080/10962247.2017.1280560

[77] M. Gao, S.Y. Pan, W.C. Chen, P.C. Chiang, A cross-disciplinary overview of naturally derived materials for electrochemical energy storage, Mater Today Energy 7 (2018) 58–79. https://doi.org/10.1016/j.mtener.2017.12.005

[78] G. Zhang, Y. Chen, Y. Chen, H. Guo, Activated biomass carbon made from bamboo as electrode material for supercapacitors, Mater. Res. Bul. 102 (2018) 391–398. https://doi.org/10.1016/j.materresbull.2018.03.006

[79] H. Sun, W. He, C. Zong, L. Lu, Template-free synthesis of renewable macroporous carbon via yeast cells for high performance supercapacitor electrode

materials. ACS Appl. Mater. Interfaces 5 (2013) 2261–2268.
https://doi.org/10.1021/am400206r

[80] Y.T. Li, Y.T. Pi, L.M. Lu, S.H. Xu, T.Z. Ren, Hierarchical porous active carbon from fallen leaves by synergy of K_2CO_3 and their supercapacitor performance, J. Power Sources 299 (2015) 519–528.
https://doi.org/10.1016/j.jpowsour.2015.09.039

[81] X.L. Su, J.R. Chen, G.P. Zheng, J.H. Yang, X.X. Guan, P. Liu, X.C. Zheng, Three-dimensional porous activated carbon derived from loofah sponge biomass for supercapacitor applications, Appl. Surf. Sci. 436 (2018) 327–336.
https://doi.org/10.1016/j.apsusc.2017.11.249

[82] K. Wang, N. Zhao, S. Lei, R. Yan, X. Tian, J. Wang, Y. Song, D. Xu, Q. Guo, L. Liu, Promising biomass based activated carbons derived from willow catkins for high performance supercapacitors, Electrochim. Acta 166 (2015) 1–11.
https://doi.org/10.1016/j.electacta.2015.03.048

[83] R. Wang, P. Wang, X. Yan, J. Lang, C. Peng, Q. Xue, Promising porous carbon derived from celtuce leaves with outstanding supercapacitance and CO_2 capture performance, ACS Appl. Mater. Interfaces 4 (2012) 5800–5806.
https://doi.org/10.1021/am302077c

[84] C. Peng, X.B. Yan, R.T. Wang, J.W. Lang, Y.J. Ou, Q.J Xue, Promising activated carbons derived from waste tea-leaves and their application in high performance supercapacitors electrodes, Electrochim. Acta 87 (2013) 401–408.
https://doi.org/10.1016/j.electacta.2012.09.082

[85] P. Veerakumar, C. Rajkumar, S.M. Chen, B. Thirumalraj, K.C. Li, Activated porous carbon supported rhenium composites as electrode materials for electrocatalytic and supercapacitor applications, Electrochim. Acta 271 (2018) 433–447. https://doi.org/10.1016/j.electacta.2018.03.165

[86] X. Zhu, S. Yu, K. Xu, Y. Zhang, L. Zhang, G. Lou, Y. Wu, E. Zhu, H. Chen, Z. Shen, B. Bao, S. Fu. Sustainable activated carbons from dead ginkgo leaves for supercapacitor electrode active materials, Chem. Eng. Sci. 181 (2018) 36–45.
https://doi.org/10.1016/j.ces.2018.02.004

[87] D. Bhattacharjya, J.S. Yu, Activated carbon made from cow dung as electrode material for electrochemical double layer capacitor, J Power Sources 262 (2014) 224–231. https://doi.org/10.1016/j.jpowsour.2014.03.143

[88] J. Hou, C. Cao, F. Idrees, X. Ma, Hierarchical porous nitrogen-doped carbon nanosheets derived from silk for ultrahigh-capacity battery anodes and

supercapacitors, ACS Nano 9 (2015) 2556–2564.
https://doi.org/10.1021/nn506394r

[89]   W. Qian, F. Sun, Y. Xu, L. Qiu, C. Liu, S. Wang, F. Yan, Human hair-derived
       carbon flakes for electrochemical supercapacitors, Energy Environ Sci. 7 (2014)
       379–386. https://doi.org/10.1039/C3EE43111H

[90]   H. Feng, M. Zheng, H. Dong, Y. Xiao, H. Hu, Z. Sun, C. Long, Y. Cai, X. Zhao,
       H. Zhang, B. Lei, Y. Liu, Three-dimensional honeycomb-like hierarchically
       structured carbon for high-performance supercapacitors derived from high ash-
       content sewage sludge, J Mater Chem A. 3 (2015) 15225–15234.
       https://doi.org/10.1039/C5TA03217B

[91]   S.K. Hoekman, A. Broch, C. Robbins, Hydrothermal carbonization (HTC) of
       lignocellulosic biomass, Energy Fuels 25 (2011) 1802–1810.
       https://doi.org/10.1021/ef101745n

[92]   S. Nizamuddina, H.A. Baloch, G.J. Griffin, N.M. Mubarak, A.W. Bhutto, R.
       Abrod, S.A. Mazari, B.S. Alie, An overview of effect of process parameters on
       hydrothermal carbonization of biomass, Renew. Sustain. Energy. Rev. 73 (2017)
       1289–1299. https://doi.org/10.1016/j.rser.2016.12.122

[93]   Wang T, Zhai Y, Zhu Y, Li C, Zeng G. A review of the hydrothermal
       carbonization of biomass waste for hydrochar formation: Process conditions,
       fundamentals, and physicochemical properties, Renew. Sustain. Energy Rev. 90
       (2018) 223–247. https://doi.org/10.1016/j.rser.2018.03.071

[94]   L. Wei, M. Sevilla, A.B. Fuertes, R. Mokaya, G. Yushin, Hydrothermal
       carbonization of abundant renewable natural organic chemicals for high-
       performance supercapacitor electrodes, Adv Energy Mater.1 (2011) 356–361.
       https://doi.org/10.1002/aenm.201100019

[95]   C. Falco, J.M. Sieben, N. Brun, M. Sevilla, T. Mauelen, E. Morallón, D. Cazorla-
       Amorós, M.-M. Titirici, Hydrothermal carbons from hemicellulose-derived
       aqueous hydrolysis products as electrode materials for supercapacitors,
       ChemSusChem 6 (2013) 374–382. https://doi.org/10.1002/cssc.201200817

[96]   H. Wang, Z. Li, J.K. Tak, C.M.B. Holt, X. Tan, Z. Xu, B.S. Amirkhiz, D.
       Harfield, A. Anyia, T. Stephenson, D. Mitlin, Supercapacitors based on carbons
       with tuned porosity derived from paper pulp mill sludge biowaste, Carbon 57
       (2013) 317–328. https://doi.org/10.1016/j.carbon.2013.01.079

[97]   L. Zhao, L.Z. Fan, M.Q. Zhou, H. Guan, S. Qiao, M. Antonietti, M.-M. Titirici,
       Nitrogen-containing hydrothermal carbons with superior performance in

supercapacitors, Adv. Mater. 22 (2010) 5202–5206.
https://doi.org/10.1002/adma.201002647

[98] C. Long, X. Chen, L. Jiang, L. Zhi, Z. Fan, Porous layer-stacking carbon derived from in-built template in biomass for high volumetric performance supercapacitors, Nano Energy 12 (2015) 141–151.
https://doi.org/10.1016/j.nanoen.2014.12.014

[99] W. Tian, Q. Gao, Y. Tan, K. Yang, L. Zhu, C. Yang, H. Zhang, Bio-inspired beehive-like hierarchical nanoporous carbon derived from bamboo based industrial by-product as a high performance supercapacitor electrode material, J. Mater. Chem. A 3 (2015) 5656–5664. https://doi.org/10.1039/C4TA06620K

[100] Sevilla M, Gu W, Falco C, M.M. Titirici, A.B. Fuertes, G. Yushin, Hydrothermal synthesis of microalgae-derived microporous carbons for electrochemical capacitors. J Power Sources 267 (2014) 26–32.
https://doi.org/10.1016/j.jpowsour.2014.05.046

[101] X.L. Wu, T. Wen, H.L. Guo, S. Yang, X. Wang, A.-W. Xu, Biomass-derived sponge-like carbonaceous hydrogels and aerogels for supercapacitors, ACS Nano 7 (2013) 3589–3597. https://doi.org/10.1021/nn400566d

[102] H. Wang, Z. Xu, A. Kohandehghan, Z. Li, K. Cui, X. Tan, T.J. Stephenson, C.K. Kingondu, C.M. B. Holt, B.C. Olsen, J.K. Tak, D. Harfield, A.O. Anyia, D. Mitlin, Interconnected carbon nanosheets derived from hemp for ultrafast supercapacitors with high energy, ACS Nano 7 (2013) 5131–5141.
https://doi.org/10.1021/nn400731g

[103] Y. Gong, L. Xie, H. Li, Y. Wang, Sustainable and scalable production of monodisperse and highly uniform colloidal carbonaceous spheres using sodium polyacrylate as the dispersant, Chem. Commun. 50 (2014) 2633–12636.
https://doi.org/10.1039/C4CC04998E

[104] Y. Wu, J.P. Cao, X.Y. Zhao, Z.Q. Hao, Q.Q. Zhuang, J.S. Zhu, X.Y. Wang, X.Y. Wei, Preparation of porous carbons by hydrothermal carbonization and KOH activation of lignite and their performance for electric double layer capacitor, Electrochim. Acta 252 (2017) 397–407.
https://doi.org/10.1016/j.electacta.2017.08.176

[105] Z.G. Liu, F.S. Zhang, J.Z. Wu, Characterization and application of chars produced from pinewood pyrolysis and hydrothermal treatment, Fuel 89 (2010) 510–514.
https://doi.org/10.1016/j.fuel.2009.08.042

[106] B.R. Selvi, D. Jagadeesan, B.S. Suma, G. Nagashankar, M. Arif, K. Balasubramanyam, M. Eswaramoorthy, T.K. Kundu, Intrinsically fluorescent

carbon nanospheres as a nuclear targeting vector: Delivery of membrane impermeable molecule to modulate gene expression in vivo, Nano Lett. 8 (2008) 3182–3188. https://doi.org/10.1021/nl801503m

[107] S. Liu, Y. Liang, W. Zhou, W. Hu, H. Dong, M. Zheng, H. Hu, B. Lei, Y. Xiao, Y. Liu, Large-scale synthesis of porous carbon via one-step $CuCl_2$ activation of rape pollen for high performance supercapacitors, J. Mater. Chem. A 6 (2018) 12046–12055. https://doi.org/10.1039/C8TA02838A

[108] K.H. Adolfsson, C.F. Lin, M. Hakkarainen, Microwave assisted hydrothermal carbonization and solid state post-modification of carbonized polypropylene, ACS Sustainable Chem. Eng. 6 (2018) 11105−11114. https://doi.org/10.1021/acssuschemeng.8b02580

[109] Z. Hu, S. Jin, W. Lu, S. Tang, C. Guo, Y. Lu, R. Zhang, Y. Liu, M. Jin, Effect of carbonization temperature on microwave absorbing properties of polyacrylonitrile-based carbon fibers, Fuller. Nanotube. Carbon 25 (2017) 637−641. https://doi.org/10.1080/1536383X.2017.1372751

[110] Y. Cheng, B. Li, Y. Huang, Y. Wang, J. Chen, D. Wei, Y. Feng, D. Jia, Y. Zhou, Molten salt synthesis of nitrogen and oxygen enriched hierarchically porous carbons derived from biomass via rapid microwave carbonization for high voltage supercapacitors, Appl. Surf. Sci. 439 (2018) 712–723. https://doi.org/10.1016/j.apsusc.2018.01.006

[111] J.S. Lee, R.T. Mayes, H. Luo, S. Dai, Ionothermal carbonization of sugars in a protic ionic liquid under ambient conditions, Carbon 48 (2010) 3364–3368. https://doi.org/10.1016/j.carbon.2010.05.027

[112] E. Raymundo-Pinero, M. Cadek, F. Beguin, Tuning carbon materials for supercapacitors by direct pyrolysis of seaweeds, Adv. Funct. Mater. 19 (2009) 1032–1039. https://doi.org/10.1002/adfm.200801057

[113] Y.D. Chen, M.J. Huang, B. Huang, X.R. Chen, Mesoporous activated carbon from inherently potassium rich pokeweed by in situ self-activation and its use for phenol removal, J. Anal. Appl. Pyrolysis 98 (2012) 159–165. https://doi.org/10.1016/j.jaap.2012.09.011

[114] M. Biswal, A. Banerjee, M. Deo, S. Ogale, From dead leaves to high energy density supercapacitors, Energy Environ. Sci. 6 (2013) 1249–1259. https://doi.org/10.1039/c3ee22325f

[115] D. Xin-hui, C. Srinivasakannan, P. Jin-hui, Z. Li-bo, Z. Zheng-yong, Comparison of activated carbon prepared from Jatropha hull by conventional heating and

microwave heating, Biomass Bioenergy 35 (2011) 3920–3926.
https://doi.org/10.1016/j.biombioe.2011.06.010

[116] Y.M. Chen, S. Ji, H. Wang, V. Linkov, R.F. Wang, Synthesis of porous nitrogen and sulfur co-doped carbon beehive in a high-melting-point molten salt medium for improved catalytic activity toward oxygen reduction reaction, Int. J. Hydrogen Energy 43 (2018) 5124–5132. https://doi.org/10.1016/j.ijhydene.2018.01.095

[117] F. Yang, L.L. Sun, W.L. Xie, Q. Jiang, Y. Gao, W. Zhang, Y. Zhang, Nitrogen-functionalization biochars derived from wheat straws via molten salt synthesis: An efficient adsorbent for atrazine removal, Sci. Total Environ., 607 (2017) 1391–1399. https://doi.org/10.1016/j.scitotenv.2017.07.020

[118] H.S. Shang, Y.J. Lu, F. Zhao, C. Chao, B. Zhang, H.S. Zhang, Preparing high surface area porous carbon from biomass by carbonization in a molten salt medium, RSC Adv. 5 (2015) 75728–75734. https://doi.org/10.1039/C5RA12406A

[119] C.J. Wang, D.P. Wu, H.J. Wang, Z.Y. Gao, F. Xu, K. Jiang, A green and scalable route to yield porous carbon sheets from biomass for supercapacitors with high capacity, J. Mater. Chem. A 6 (2018) 1244–1254.
https://doi.org/10.1039/C7TA07579K

[120] Z. Wang, D. Shen, C. Wu, S. Gu, State-of-the-art on the production and application of carbon nanomaterials from biomass, Green Chem. 20 (2018) 5031–5057. https://doi.org/10.1039/C8GC01748D

[121] Y. Liu, B. Huang, X. Lin, Z. Xie, Biomass-derived hierarchical porous carbons: boosting the energy density of supercapacitors via an ionothermal approach, J. Mater. Chem. A 5 (2017) 13009–13018. https://doi.org/10.1039/C7TA03639F

[122] C. Xia, S.Q. Shi, Self-activation for activated carbon from biomass: theory and parameters. Green Chem. 18 (2016) 2063–2071.
https://doi.org/10.1039/C5GC02152A

[123] K. Sun, C.Y. Leng, J.C. Jiang, Q. Bu, G.F. Lin, X.C. Lu, G.Z. Zhu, Microporous activated carbons from coconut shells produced by self-activation using the pyrolysis gases produced from them, that have an excellent electric double layer performance, New Carbon Mater. 32 (2017) 451–459.
https://doi.org/10.1016/S1872-5805(17)60134-3

[124] S.Q. Shi, C. Xia, Porositization process of carbon or carbonaceous materials. US Patent App. (2014) 14/211, 357.

[125] C. Xia, C. Kang, M.D. Patel, L. Cai, B. Gwalani, R. Banerjee, S.Q. Shi, W. Choi, Pine wood extracted activated carbon through self-activation process for high-

performance lithium-ion battery, Chemistry Select 1 (2016) 4000–4007.
https://doi.org/10.1002/slct.201600926

[126] C. Bommier, R. Xu, W. Wang, X. Wang, D. Wen, J. Lu, X. Ji, Self-activation of
cellulose: A new preparation methodology for activated carbon electrodes in
electrochemical capacitors, Nano Energy 13 (2015) 709–717.
https://doi.org/10.1016/j.nanoen.2015.03.022

[127] S. Herou, P. Schlee, A.B. Jorge, M. Titirici, Biomass-derived electrodes for
flexible supercapacitors. Curr Opin Green Sustain. Chem. 9 (2018) 18–24.
https://doi.org/10.1016/j.cogsc.2017.10.005

[128] S. Dutta, A. Bhaumik, K.C.W. Wu, Hierarchically porous carbon derived from
polymers and biomass: Effect of interconnected pores on energy applications,
Energy Environ. Sci. 7 (2014) 3574–3592. https://doi.org/10.1039/C4EE01075B

[129] Y. Liu, J. Chen, B. Cui, P. Yin, C. Zhang, Design and preparation of biomass-
derived carbon materials for supercapacitors: A review, C 4 (2018) 53.
https://doi.org/10.3390/c4040053

[130] L. Jiang, L. Sheng, Z. Fan, Biomass-derived carbon materials with structural
diversities and their applications in energy storage, Sci. Chin. Mater. 61 (2018)
133–158. https://doi.org/10.1007/s40843-017-9169-4

[131] R.J. Mo, Y. Zhao, M. Wu, H.M. Xiao, S. Kuga, Y. Huang, J.P. Li, S.Y. Fu,
Activated carbon from nitrogen rich watermelon rind for high-performance
supercapacitors, RSC Adv. 6 (2016) 59333–59342.
https://doi.org/10.1039/C6RA10719B

[132] Y.Y. Wang, B.H. Hou, H.Y. Lü, C.L. Lü, X.L. Wu, Hierarchically porous N-
doped carbon nanosheets derived from grapefruit peels for high-performance
supercapacitors, Chemistry Select 1 (2016) 1441–1447.
https://doi.org/10.1002/slct.201600133

[133] K. Chaitra, R.T. Vinny, P. Sivaraman, N. Reddy, C. Hu, K. Venkatesh, C.S.
Vivek, N. Nagaraju, N. Kathyayini, KOH activated carbon derived from biomass-
banana fibers as an efficient negative electrode in high performance asymmetric
supercapacitor, J. Energy Chem. 26 (2017) 56–62.
https://doi.org/10.1016/j.jechem.2016.07.003

[134] T.E. Rufford, D. Hulicova-Jurcakova, Z. Zhu, G.Q. Lu, Nanoporous carbon
electrode from waste coffee beans for high performance supercapacitors,
Electrochem. Commun. 10 (2008) 1594–1597.
https://doi.org/10.1016/j.elecom.2008.08.022

[135] Y. Lv, L. Gan, M. Liu, W. Xiong, Z. Xu, D. Zhu, D.S. Wright, A self-template synthesis of hierarchical porous carbon foams based on banana peel for supercapacitor electrodes, J. Power Sources 209 (2012) 152–157. https://doi.org/10.1016/j.jpowsour.2012.02.089

[136] J. Chang, Z. Gao, X. Wang, D. Wu, F. Xu, X. Wang, Y. Guo, K. Jiang, Activated porous carbon prepared from paulownia flower for high performance supercapacitor electrodes, Electrochim. Acta 157 (2015) 290–298. https://doi.org/10.1016/j.electacta.2014.12.169

[137] Y. Fan, X. Yang, B. Zhu, P.F. Liu, H.T. Lu, Micro-mesoporous carbon spheres derived from carrageenan as electrode material for supercapacitors, J. Power Sources 268 (2014) 584–590. https://doi.org/10.1016/j.jpowsour.2014.06.100

[138] L. Zhang, T. You, T. Zhou, X. Zhou, F. Xu, Interconnected hierarchical porous carbon from lignin-derived byproducts of bioethanol production for ultra-high performance supercapacitors, ACS Appl. Mater. Interfaces 8 (2016) 13918–13925. https://doi.org/10.1021/acsami.6b02774

[139] M. Karnan, K. Subramani, P.K. Srividhya, M. Sathish, Electrochemical studies on corncob derived activated porous carbon for supercapacitors application in aqueous and non-aqueous electrolytes, Electrochim. Acta 228 (2017) 586–596. https://doi.org/10.1016/j.electacta.2017.01.095

[140] L. Peng, Y. Liang, H. Dong, H. Hu, X. Zhao, Y. Cai, Y. Xiao, Y. Liu, M. Zheng, Super-hierarchical porous carbons derived from mixed biomass wastes by a stepwise removal strategy for high-performance supercapacitors, J. Power Sources 377 (2018) 151–160. https://doi.org/10.1016/j.jpowsour.2017.12.012

Biomass Based Energy Storage Materials
Materials Research Foundations **78** (2020) 91-110

Materials Research Forum LLC
https://doi.org/10.21741/9781644900871-4

Chapter 4

Lignin-Derived Materials for Energy Storage

Paul Thomas [1], Nelson Pynadathu Rumjit [1], Shivani Garg [2], Chin Wei Lai [1]*,
Mohd Rafie Bin Johan [1]

[1] Nanotechnology & Catalysis Research Centre, Institute for Advanced Studies (IAS), University of Malaya (UM), Level 3, Block A, 50603 Kuala Lumpur, Malaysia

[2] Institute of Environmental Studies, Kurukshetra University, Kurukshetra 136119, Haryana, India

*cwlai@um.edu.my

Abstract

Lignin is an abundant by-product derived from biorefinery, paper and pulp industry and it is one of the most inexpensive natural biopolymer. Although lignin has been used for broad applications, the suitability of lignin for energy storage has not been explored in detail. Lignin suitability is mainly utilized as binders, electrodes for batteries and supercapacitors. The application of lignin in energy storage devices enhanced the performance of energy storage devices and also makes it eco-friendly and cheaper. This chapter focuses on the application of lignin towards fabrication and replacement of toxic and synthetic compounds with emphasis on batteries, supercapacitors and other energy storage devices.

Keywords

Lignin, Supercapacitor, Energy Storage, Composite Materials

Contents

1. Introduction 92

2. Lignin isolation process 94

3. Lignin carbon fibres 94

    3.1   Activation techniques 95

    3.2   Lignin- Lignin blends 96

    3.3   Lignin-Cellulose blends 96

3.4   Fractionation 96

3.5   Reinforcement..97

3.6   Chemical modification..97

3.7   New lignin types 97

4.   Lignin-derived porous carbon..**98**

5.   Challenges with graphite-based electrodes **98**

6.   Lignin for electrochemical applications **99**

6.1   Lithium-ion batteries 99

6.2   Electrochemical double layer capacitors 99

6.3   Electrochemical pseudocapacitors..100

6.4   Sodium –ion batteries 101

6.5   Lignin as binder 101

Conclusion and Perspectives...**102**

Acknowledgements..**103**

References **103**

1.   Introduction

Considering the abundance of polymers present in nature, lignin holds the second position after cellulose [1]. A vast quantity of lignin is produced worldwide as a by-product from various industries such as wood processing, biofuel and alcohol industry and it is estimated to be 70 million tons annually [2,3]. The production of lignin wastes will increase due to new initiatives set up to replace the hydrocarbon fuel sources and chemical products by bioresources. Biorefineries considered lignin as a waste product priority, and attention was focused only towards the valorisation of hemicellulose and cellulose [4]. The commonly followed harsh processing techniques in biorefineries led to highly heterogeneous, complex and polydispersed lignin which acts as a barrier to its widescale economic valorisation. However, recent development focusing on lignin as primary valorisation without imperiling the carbohydrate fraction could enhance the characteristics of derived lignin and its transformation into valuable by-products [5,6]. Lignin as an abundant natural biopolymer with high aromatic and carbon content could contribute as a significant industrial precursor for the production of both structural (carbon black, carbon fibre) and functional (catalyst and electrodes) carbon materials [7,8].

The lignin-derived carbon materials gained attention in recent years and are proliferating in the field of energy and environment. The limited availability of fossil fuel creates a severe issue for humanity. The demand for high capacity, budget-friendly and reusable energy storage devices are rising in today's energy intensive world. Future energy generation and storage are not only cost-effective but also need to be sustainable. Among various commercially available energy storage devices, rechargeable lithium-ion is the most attractive option as it inherits superior energy density with desirable dischargeable characteristics. Lithium-ion batteries have been extensively utilised in various high-end electronic devices such as hybrid electric vehicles, computer laptops, mobiles etc. [9]. However, the expensive production cost, safety and eventual safe disposals are significant constraints in the manufacturing of energy storage devices. Numerous efforts have been carried out to integrate bio-derived polymers such as lignin and cellulose in the scientific community. Fig. 1 [10] exhibits average chemical composition of lignocellulosic materials and a brief scheme of primary inhibitory compounds formation. The objective of this chapter is focused on the conversion of lignin to value-added by-products with emphasis on energy storage applications.

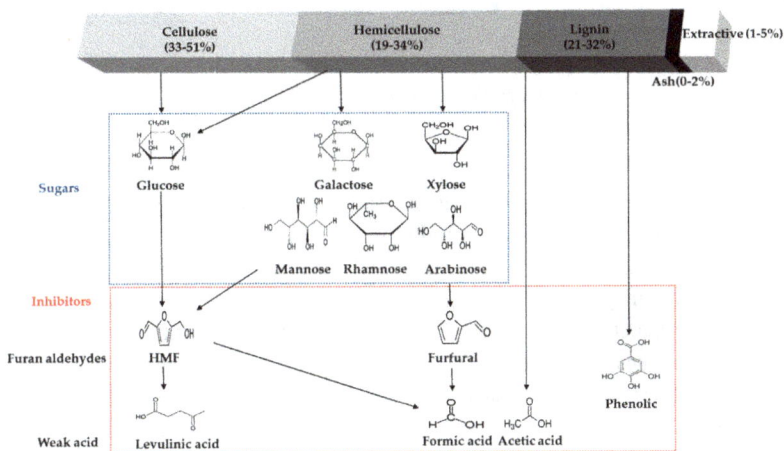

Fig. 1: The average chemical composition of lignocellulosic materials and a brief scheme of main inhibitory compounds formation [10].

Biomass Based Energy Storage Materials   Materials Research Forum LLC
Materials Research Foundations **78** (2020) 91-110   https://doi.org/10.21741/9781644900871-4

Fig. 2: *Overview of potential applications of Lignin from microscale to nanoscale [13].*

2. Lignin isolation process

The primary source of lignin is plants. Plants are composed of lignin, cellulose and hemicellulose. The cellulose and hemicellulose are the major constituents (80%), while lignin constitutes 30%. Lignin is covalently bound to the polysaccharide cell wall constituents of both polyose and cellulose. Lignin is segregated from other polymers using various separation techniques [11]. The sulphite and sulphate processes are the dominant techniques in isolation and production of lignin commercially. In these isolation techniques, lignosulfonates and kraft lignin are the main by-products derived through these separation techniques from wood. During the sulphate interaction, kraft lignin derived from the remaining compounds under strong alkaline condition breaks lignin into smaller chains [12]. In lignosulfonate, extraction is carried out by a similar approach by replacing alkaline condition with an acidic medium. There are numerous techniques to isolate ionic liquid lignin, organosolv lignin and soda lignin in laboratory

Biomass Based Energy Storage Materials | Materials Research Forum LLC
Materials Research Foundations **78** (2020) 91-110 | https://doi.org/10.21741/9781644900871-4

scale and pilot plants. Organosolv lignin is extracted using alcohols or acidic solutions, soda lignin is obtained using alkaline solution, and ionic liquids are processed using salts. The commercial-scale production of these lignins is still challenging, and more research needs to be carried out to increase its application. Fig. 2 [13] demonstrates potential applications of lignin from microscale to nanoscale.

3. Lignin carbon fibres

Carbon fibre is the most valuable by-product from lignin and gained the attraction of researchers and the scientific community. The light-weight materials like carbon fibre inherit superior mechanical characteristics which have an extensive use in a wide variety of applications from aerospace, automotive, energy and other industries [7,14]. Polyacrylonitrile is the primary precursor for the production of carbon fibre [15]. However, the major drawback in the production of carbon fibre from polyacrylonitrile is the cost of precursor polyacrylonitrile which is approximately \$33/kg [16]. Due to the bulk availability, low cost and bio-renewable nature, lignin will be an attractive substitute to replace polyacrylonitrile.

Carbon fibre is mainly categorised based on applications in general purpose and high-performance carbon fibre. High-performance carbon fibre is mainly used in the field of automobile manufacturing, sporting and reinforcing composites, whereas the general-purpose carbon fibre is used as activated carbon fibre in electrodes for electrochemical applications and as the catalyst for environmental applications. A significant challenge associated with the processing of lignin-based carbon fibre is the spinnability of lignins. The properties and processing of lignins are influenced by various factors such as molecular weight distribution, molecular and atomic bonds, molecular chain configuration and conformation and degree/order and disorder in the molecular chains [17]. Hardwood lignin is more suitable as primary raw material than softwood lignin as it cannot be melt spun.

3.1 Activation techniques

Physical and chemical activation techniques are the two main techniques employed to obtain activate lignin carbon. Physical activation technique is carried out in the presence of gas including steam, carbon dioxide or air. Air activation is usually carried out at a low temperature below 500 °C. Steam is also a leading activation agent for precursor due to low cost and its ability to remove by-products [18]. Carbon dioxide-based activation is carried out at high temperature. The carbon dioxide-based activation develops mesopores which is suitable for broad energy storage applications. In chemical activation, agents such as KOH, $ZnCl_2$ and H_3PO_4 have been utilised in the activation of lignin-derived

Biomass Based Energy Storage Materials   Materials Research Forum LLC
Materials Research Foundations **78** (2020) 91-110   https://doi.org/10.21741/9781644900871-4

carbon. KOH is widely used for activating the lignin-derived carbon to obtain more hierarchical pores with high specific surface area.

3.2   Lignin- Lignin blends

To improve the characteristics of pure lignin, carbon fibre extracted from hardwood blending with herbaceous lignin has been proposed recently [19]. The yellow poplar lignin blended with switch grass lignin enhances the thermostabilising performance and mechanical characteristics that also prevent the fusion of fibres. The derived carbon fibre has tensile strength of 230-750 MPa, Youngs modules 30.4-41.8 and diameter in the range of 16-31μm. This proposed approach significantly decreases the production and processing time, helps in producing renewable carbon fibre and enhances mechanical characteristics.

3.3   Lignin-Cellulose blends

The addition of nanocrystalline cellulose develops lignin-based carbon fibre with a similar cross-section through direct carbonisation without performing oxidative and thermal stabilisation treatment [20]. This processing technique develops unique interconnected carbon mats with enhanced electrical conductivity by preventing the fusion of fibres. This processing approach is economically profitable by reducing energy consumption and processing time in production.

3.4   Fractionation

Softwood kraft lignin was used as a precursor to derive smooth and solid carbon fibre [21]. The solid and smooth carbon fibre is processed by the direct spinning of the permeate acquired by ultrafiltration of black liquor derived through ceramic membrane pursued by carbonisation and oxidative stabilisation treatment. The processed solid and smooth carbon fibre yielded a rich carbon content of 97%. Fig. 3 exhibits scanning electron microscopy images of carbon fibres derived from yellow poplar, switchgrass and lignin blends. Recently enzyme mediator system was utilised to fabricate high-quality lignin-based carbon fibre [4].

Fractionation technique enhances the spinnability of lignin and produces the finer lignin carbon fibre. The increase in elastic modulus of lignin carbon fibre is related to the increase in molecular weight of lignin. Also, the increase in molecular weight leads to low polydispersity index which in turns increases the crystallite structure of lignin-based carbon fibre [22]. This modification approach paved new opportunities in the processing of high-quality lignin-based carbon fibres.

3.5   Reinforcement

The grafting of lignin-derived carbon fibre with carbon nanotubes develops well-oriented carbon fibres with excellent mechanical characteristics [23]. The tensile strength was enhanced to 289.3 MPa; however, the presence of voids appeared in the carbonised lignin fibres which arise due to the breakage of chemical links between lignin and carbon nanotubes.

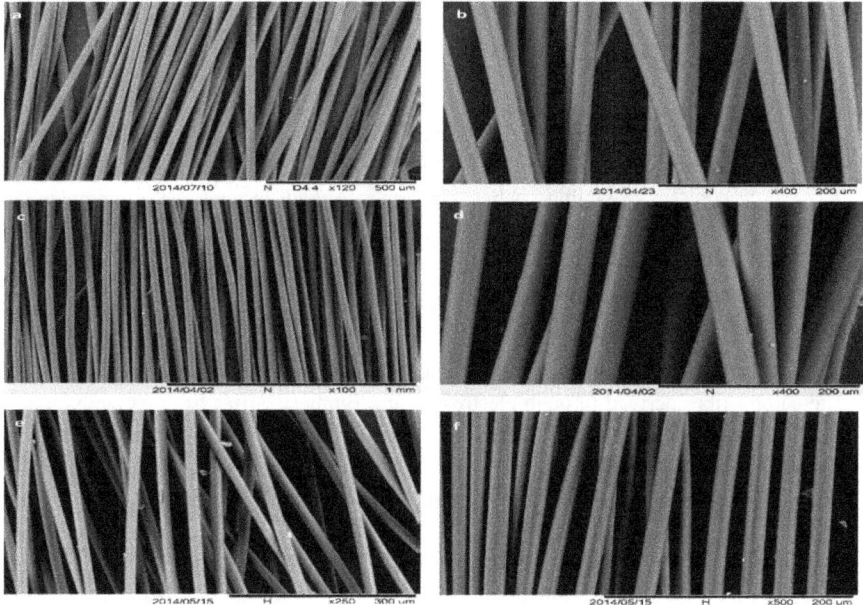

Fig. 3:Scanning electron microscopy images of carbon fibers derived from yellow poplar and switch grass and lignin blends: (a, b) 50% yellow poplar:50% switch grass thermostabilized at heating rate of 0.05 °C min⁻¹, (c,d) 50% yellow poplar:50% switch grass thermostabilized at heating rate of 0.5 °C min⁻¹, (e) 75% yellow poplar:25% switchgrass thermostabilized at heating rate of 0.05 °C min⁻¹, and (f) 85% yellow poplar:15% switchgrassthermostabilized at heating rate of 0.05 °C min⁻¹ [19].

3.6   Chemical modification

The influence of acetylation treatment was observed on the fabrication of carbon fibre from precursor corn stover lignin [24]. Fractionation with methanol has been employed to remove the impurities present in the lignin, and it is also able to remove molecules with high melting points. The two-step acetylation of methanol fractionated lignin enhanced

Biomass Based Energy Storage Materials   Materials Research Forum LLC
Materials Research Foundations **78** (2020) 91-110   https://doi.org/10.21741/9781644900871-4

the lignin mobility and thermal stability of precursor during the melt spinning process. Similarly, the iodine pre-treatment of lignin-based precursor carbon fibre improves the thermal stability and reduces the energy consumption of the process [25]. The carbon fibre derived from iodine pre-treated lignin carbon fibre resulted in high modulus, tensile strength and yield.

3.7   New lignin types

Recently, a new category of lignin poly- (caffeyl alcohol) has been extracted from the vanilla orchid seeds which have been utilised for the production of carbon fibres with additional blending or modification with polymers [26]. Poly- (caffeyl alcohol) composed of polyaromatic networks, derived from caffeyl alcohol monomers is linked head to tail into benzodioxane chains. Studies regarding the utilisation of poly- (caffeyl alcohol) derived carbon fibre exhibits better performance and characteristics compared to commercial, and kraft derived carbon fibre.

4.   Lignin-derived porous carbon

Lignin-derived porous carbon arises as a desirable option for the fabrication of electrode, and the presence of functional groups could enhance electrochemical performance [27]. The carbon fibre derived through electrospinning synthesis arises as an attractive option [28,29]. [28,29]. The activated carbon fibre synthesised through KOH and NaOH from alkali lignin exhibits high specific surface area with an excellent specific capacitance of 344 F^{g-1}. Another interesting synthesis approach to prepare lignin-derived porous carbon is through a template free or template method [30–32]. Mesoporous carbon was synthesised using surfactant Pluronic F127, a triblock copolymer as the template. Subsequent physical activation uses carbon dioxide followed by chemical activation using KOH. The carbon dioxide activated mesoporous carbon demonstrated a capacitance of 102.3 F^{g-1} and chemically activated (KOH) demonstrated 91.7 F^{g-1}[33].

5.   Challenges with graphite-based electrodes

The utilisation of graphite as electrodes has several drawbacks such as the production of the conventional graphitic electrode which requires significant processing that involves current collectors and binders which make the entire process expensive and time-consuming. Furthermore, the electrolyte solution may promote exfoliation and leads to an unstable solid electrolyte interface that finally results in battery failure [34]. These limitations result in developing high capacity and cheaper materials that could enhance cycling stability. Over the past few decades, many research investigations have been carried out in developing alternative carbons that can insert lithium ions at higher capacity when compared to graphite. Several studies have been carried out to prepare new disordered carbon materials; however, carbon material processing requires expensive chemicals, and it has limited economic viability. The high level of microporosity in the electrode makes it more robust and least affected by electrolyte interactions [35]. Lignin-derived carbon fibres entail turbostratic disorder, and microscale porosity has some distinct advantages over graphite. Besides, the usage of lignin as the electrode material is cheaper compared to graphite and other composites.

6.   Lignin for electrochemical applications

Lignin-derived carbon has excellent potential in electrochemical applications due to its inherited characteristics such as 1) Less cost, 2) Superior electrical conductivity, 3) Porous structure, 4) High specific surface area and 5) Corrosive resistant. The lignin-based electrochemical application is discussed below.

6.1   Lithium-ion batteries

The nitrogen-containing hollow carbon nanospheres derived from lignosulfonate-polyethylaniline polymer demonstrated outstanding electrochemical characteristics for lithium-ion batteries [36]. The electrode demonstrated an excellent coulombic efficiency of 94% with superior cycling performance. These hollow carbon nanospheres can act as next-generation energy storage materials for Lithium-ion batteries. Similar studies were carried out by Su Xi Wang, et al. [37] who successfully fused carbon fibrous mat which demonstrated outstanding specific capacitance of 445 mA h/g at the current density of 30 mA/g with excellent cyclic capabilities. The lignin-derived nanocarbon materials suit energy storage applications. Lignin-derived porous carbon acts as an electrode for lithium-sulfur battery prepared by KOH based activation at 700°C [38]. The derived carbon is porous in nature with high specific surface area, and it also demonstrated excellent electrochemical performance with a charge capacity of 791.6 mA h/g and discharge capacity of 1241 mA h/g.

6.2   Electrochemical double layer capacitors

The mesoporous lignin-derived electrode used a soft template method followed by carbonisation, and chemical activation using KOH for electrochemical double layer capacitor demonstrated superior electrochemical performance [33]. The chemical activation enhances the capacitance to 102.3 F/g when compared to physical activation. H. Li and colleagues [39] synthesised high-performance monolithic carbon electrode from naturally available lignin through dual template approach. The carbon electrode exhibits excellent areal 3 F/cm^2 and superior volumetric capacitance 97.1 F/cm^3

Biomass Based Energy Storage Materials   Materials Research Forum LLC
Materials Research Foundations **78** (2020) 91-110   https://doi.org/10.21741/9781644900871-4

respectively for a mass loading of 14.4 mg/cm^2. It demonstrates excellent cycling capabilities with 95% retention. The carbon fibre-based fabrication of flexible and binder less electrode demonstrates superior energy and power density (10 Wh/kg and 61 kW/kg) with excellent cycling performance. Phosphorous containing lignin-derived carbon fibres exhibits a much more stable performance in a wide potential window in aqueous electrolytes Na_2SO_4, H_2SO_4 and NaOH [40]. The phosphorous-containing lignin-derived carbon fibres exhibit electrical capacity (150-200 F/g) which make them a cost-effective electrode for energy storage [41,42].

Kraft lignin-derived electrospun carbon nanofiber based supercapacitor exhibits excellent electrochemical performance [43]. The derived carbon nanofiber entails lignin (70%) with an average diameter (100 nm) and specific surface area (583 m^2/g). Lignin-derived carbon nanofiber mats exhibit superior electrochemical characteristics such as binder-free- or free-standing electrode for supercapacitor applications which demonstrated gravimetric capacitance of 50 F/g and 64 F/g at the current density of 2000 mA/g and 400 mA/g in 6M KOH aqueous electrolyte with better cycling performance. Similar studies were carried out by Sixiao H, and team [28] who derived porous submicron activated carbon fibre from lignin through electrospinning followed by annealing at 850°C. The chemical activation using NaOH or KOH develops hydrophilic conditions and is readily washable with water. The activated carbon fibre exhibited a gravimetric capacitance of 344 F/g for mass loading of 1.8 mg and maintained capacitance of 196 F/g for a high mass loading of 10 mg. The fabricated supercapacitor exhibits outstanding cycling capability and low equivalent series resistance in aqueous electrolytes. The two-dimensional cellulose nanosheet supercapacitor electrode derived from alkali kraft lignin demonstrated excellent electrochemical performance with superior conductivity (2.5 S/cm) [44]. The electrode exhibits an outstanding capacitance of 281 F/g with excellent capacitance retention (91%) and cycling performance at a current density of 0.5 A/g in 1 M H_2SO_4.

6.3   Electrochemical pseudocapacitors

Pseudocapacitors along with polymers such as polythiophene, polypyrrole and polyaniline with transition metal oxides such as Fe_3O_4, MnO_2 and RuO_2 are used as redox capacitor electrodes. Transition metal oxides exhibit better electrochemical performance; however, the scarcity, availability and processing cost limit their applications in a commercial scale. On the contrary, the availability and characteristics make conducting polymer more attractive options. However, the conducting polymers have certain drawbacks such as poor cycling stability and poor rate capabilities when compared to metal oxides [45]. Several studies reported that the combination of lignin and carbon powder could enhance electrochemical performance. The experimental investigation carried out on the interpenetrating polymer network between lignin and polypyrrole demonstrated outstanding electrochemical characteristics [46].

Presence of Quinone groups in lignosulfonates and lignin is employed as proton, electron storage and exchange during redox cycling. During the initial voltammetric cycles, aromatic methoxy groups present in the lignin gets transformed into hydroxyl groups [47]. The addition of anthraquinone sulfonate as dopants in polypyrrole based lignin composites enhanced the energy storage capabilities [48]. Lignin such as alkali lignin is not employed commonly due to its low solubility in organic acids which limits its application in energy storage systems. Ternary structured composites such as cotton/LS/PPy, polypyrrole/phosphomolybdic acid/lignosulfonates and graphene oxide/polyaniline/lignosulfonate has been synthesised as electrodes, and electrochemical analysis was conducted [49,50]. The lignosulfonate acted as a morphological controlling agent and surfactant of polyaniline whereas the carbon spheres act as spacers that prevent aggression between nanosheets. The electrodes exhibited outstanding electrochemical characteristics with specific capacitances of 304 F g^{-1} and 266 F g^{-1} respectively.

6.4 Sodium –ion batteries

In recent years sodium ion batteries have gained popularity as a next-generation future energy storage device owing to the availability of sodium. Recently, with hardwood as the precursor, it is used to derive biobased carbon nanofibrous web and was successfully fabricated which demonstrated outstanding electrochemical performance with an excellent reversible capacity of 292.6 mA h/g at the current density of 20 mA/g [51]. Availability of precursors and green synthesis approach makes carbon nanofibrous web an attractive option for future sodium-ion batteries.

6.5 Lignin as binder

Lignin is an essential constituent of lignocellulosic biomass materials that inherits a predominant aromatic architecture. Lignin acts as a backbone providing adequate stiffness and strength to cell walls. Lignin also constitutes rich carbon content which is approximately 60 wt%, and it is obtainable as a by-product from various industries such as biorefinery, pulp and paper industry. Due to its bulk availability as a waste by-product from industries, lignin elevated as an attractive substitute for the various constituents in the lithium-ion battery. In addition to its bulk availability, it also inherits properties such as high hardness, low density, heat resistant, humidity and friction [52]. Lignin has been extensively utilised in preparing anode materials for lithium-ion batteries; however, utilisation of lignin as a binder material for lithium-ion battery has not been much explored.

Table 1: *Reports of lignin-based composites for energy storage devices.*

Lignin / Composite	Specific Capacitance (F g^{-1})	Electrolyte	References
Poly(quinone-amine)/nanocarbon	144	0.1 M HClO$_4$	[55]
Solvent lignin (Aniline)	336	6M KOH	[56]
Polypyrrole/alkali lignin	377.2	1.0 M H$_2$SO$_4$	[47]
Polypyrrole/ lignosulfonates/ cotton	304	2.0 M NaCl	[49]
Lignin	124	6M KOH	[57]
Polypyrrole/ lignosulfonates	312	0.5 M H$_2$SO$_4$	[58]
Hardwood Kraft Lignin	102	6M KOH	[33]
Polypyrrole/ lignosulfonates	480	0.1 M HNO$_3$	[59]
S/O doped Softwood Lignin	231	EMI-BF$_4$	[60]
graphene oxide/polyaniline/lignosulfonate	266	6M KOH	[50]
Alkali Lignin	286	6M KOH	[61]
Alcell Lignin	116	1 M H$_2$SO$_4$	[62]

Huiran Lu and his colleagues [53] studied the suitability of lignin as the binder material for both graphite and LiFePO$_4$ electrodes. To improve the elasticity of electrodes, acetone has been utilised to dissolve lignin with 5% polyethylene glycol to this solution of graphite or LiFePO$_4$ with water and porous carbon was added to fabricate the electrodes. The fabricated electrodes were pressed to improve the contact between particles and to reduce the porosity. The fabricated electrodes exhibit superior specific capacity and better stability as a binder to embellish electrochemical characteristics. In an experimental investigation, conductive additives and conventional binders were replaced by lignin to study the high-performance anode [54]. The electrode composites demonstrated superior electrochemical characteristics with better cycling stability. In this experimental study, lignin was carbonised at a temperature of 600°C which provided excellent conductivity while also maintaining polymeric flexibility. The potential of lignin as a high-performance binder for lithium-ion battery needs to be explored further. Table 1 [55-62] represents some of the lignin-based composites that have been used as energy storage devices.

Conclusion and Perspectives

Lignin is a complex polymer which has excellent potential in the application and development of energy storage devices. The flexibility in utilisation of lignin in the broader spectrum requires greater complexities. From this perspective, lignin-derived porous activated carbon with distant morphologies combined with lower production cost will be able to substitute as electrodes for various energy storage devices. However, there are still many barriers that need to be overcome which are associated with consistency and quality of lignin. Standardised techniques for processing and separation of lignin from lignocellulosic biomass needs attention to ensure the supply of good quality lignin precursors with consistent purity and chemical composition. There lies a vast scope in the future for the fabrication of lignin-based electrodes in sustainable energy storage devices.

Acknowledgements

This research work was financially supported by the University Malaya Impact-Oriented Interdisciplinary Research Grant (No.IIRG018A-2019) and Global Collaborative Programme - SATU Joint Research Scheme (No. ST012-2019).

References

[1]   R.J.A. Gosselink, E. de Jong, B. Guran, A. Abächerli, Co-ordination network for lignin—standardisation, production and applications adapted to market requirements (EUROLIGNIN), Ind. Crops Prod. 20 (2004) 121–129. https://doi.org/10.1016/j.indcrop.2004.04.015

[2]   W.J. Liu, H. Jiang, H.Q. Yu, Thermochemical conversion of lignin to functional materials: a review and future directions, Green Chem. 17 (2015) 4888–4907. https://doi.org/10.1039/C5GC01054C

[3]   Q. Li, S. Xie, W.K. Serem, M.T. Naik, L. Liu, J.S. Yuan, Quality carbon fibers from fractionated lignin, Green Chem. 19 (2017) 1628–1634. https://doi.org/10.1039/C6GC03555H

[4]   A.G. Vishtal, A. Kraslawski, Challenges in industrial applications of technical lignins, Bioresources 6 (2011) 3547–3568.

[5] R. Rinaldi, R. Jastrzebski, M.T. Clough, J. Ralph, M. Kennema, P.C.A. Bruijnincx, B.M. Weckhuysen, Paving the way for lignin valorisation: recent advances in bioengineering, biorefining and catalysis, Angew. Chem. Int. Ed. 55 (2016) 8164–8215. https://doi.org/10.1002/anie.201510351

[6] T. Renders, S. Van den Bosch, S.F. Koelewijn, W. Schutyser, B.F. Sels, Lignin-first biomass fractionation: the advent of active stabilisation strategies, Energy Environ. Sci. 10 (2017) 1551–1557. https://doi.org/10.1039/C7EE01298E

[7] H. Mainka, O. Täger, E. Körner, L. Hilfert, S. Busse, F.T. Edelmann, A.S. Herrmann, Lignin – an alternative precursor for sustainable and cost-effective automotive carbon fiber, J. Mater. Res. Technol. 4 (2015) 283–296. https://doi.org/10.1016/j.jmrt.2015.03.004

[8] F.H.M. Graichen, W.J. Grigsby, S.J. Hill, L.G. Raymond, M. Sanglard, D.A. Smith, G.J. Thorlby, K.M. Torr, J.M. Warnes, Yes, we can make money out of lignin and other bio-based resources, Ind. Crops Prod. 106 (2017) 74–85. https://doi.org/10.1016/j.indcrop.2016.10.036

[9] M. Armand, J.M. Tarascon, Building better batteries, Nature 451 (2008) 652–657. https://doi.org/10.1038/451652a

[10] D. Kim, Kim, Daehwan, Physico-chemical conversion of lignocellulose: Inhibitor effects and detoxification strategies: A mini review, Molecules 23 (2018) 309. https://doi.org/10.3390/molecules23020309

[11] J.L. Espinoza-Acosta, P.I. Torres-Chávez, J.L. Olmedo-Martínez, A. Vega-Rios, S. Flores-Gallardo, E.A. Zaragoza-Contreras, Lignin in storage and renewable energy applications: A review, J. Energy Chem. 27 (2018) 1422–1438. https://doi.org/10.1016/j.jechem.2018.02.015

[12] J.H. Lora, W.G. Glasser, Recent industrial applications of lignin: a sustainable alternative to nonrenewable materials, J. Polym. Environ. 10 (2002) 39–48. https://doi.org/10.1023/A:1021070006895

[13] S. Beisl, A. Friedl, A. Miltner, S. Beisl, A. Friedl, A. Miltner, Lignin from Micro-to Nanosize: Applications, Int. J. Mol. Sci. 18 (2017) 2367. https://doi.org/10.3390/ijms18112367

[14] M.M. Titirici, R.J. White, N. Brun, V.L. Budarin, D.S. Su, F. del Monte, J.H. Clark, M.J. MacLachlan, Sustainable carbon materials, Chem. Soc. Rev. 44 (2015) 250–290. https://doi.org/10.1039/C4CS00232F

[15] E. Frank, L.M. Steudle, D. Ingildeev, J.M. Spörl, M.R. Buchmeiser, Carbon fibers: Precursor systems, processing, structure, and properties, Angew. Chem. Int. Ed. 53 (2014) 5262–5298. https://doi.org/10.1002/anie.201306129

[16] D.A. Baker, T.G. Rials, Recent advances in low-cost carbon fiber manufacture from lignin, J. Appl. Polym. Sci. 130 (2013) 713–728. https://doi.org/10.1002/app.39273

[17] O. Faruk, M. Sain, Lignin in Polymer Composites, 1st ed., William Andrew, United State of America, 2015.

[18] Y. Zhai, Y. Dou, D. Zhao, P.F. Fulvio, R.T. Mayes, S. Dai, Carbon materials for chemical capacitive energy storage, Adv. Mater. 23 (2011) 4828–4850. https://doi.org/10.1002/adma.201100984

[19] O. Hosseinaei, D.P. Harper, J.J. Bozell, T.G. Rials, O. Hosseinaei, D.P. Harper, J.J. Bozell, T.G. Rials, Improving processing and performance of pure lignin carbon fibers through hardwood and herbaceous lignin blends, Int. J. Mol. Sci. 18 (2017) 1410. https://doi.org/10.3390/ijms18071410

[20] M. Cho, M. Karaaslan, S. Chowdhury, F. Ko, S. Renneckar, Skipping oxidative thermal stabilization for lignin-based carbon nanofibers, ACS Sustain. Chem. Eng. 6 (2018) 6434–6444. https://doi.org/10.1021/acssuschemeng.8b00209

[21] Y. Nordström, I. Norberg, E. Sjöholm, R. Drougge, A new softening agent for melt spinning of softwood kraft lignin, J. Appl. Polym. Sci. 129 (2013) 1274–1279. https://doi.org/10.1002/app.38795

[22] Q. Li, W.K. Serem, W. Dai, Y. Yue, M.T. Naik, S. Xie, P. Karki, L. Liu, H.J. Sue, H. Liang, F. Zhou, J.S. Yuan, Molecular weight and uniformity define the mechanical performance of lignin-based carbon fiber, J. Mater. Chem. A. 5 (2017) 12740–12746. https://doi.org/10.1039/C7TA01187C

[23] S. Wang, Z. Zhou, H. Xiang, W. Chen, E. Yin, T. Chang, M. Zhu, Reinforcement of lignin-based carbon fibers with functionalized carbon nanotubes, Compos. Sci. Technol. 128 (2016) 116–122. https://doi.org/10.1016/j.compscitech.2016.03.018

[24] W. Qu, J. Liu, Y. Xue, X. Wang, X. Bai, Potential of producing carbon fiber from biorefinery corn stover lignin with high ash content, J. Appl. Polym. Sci. 135 (2018) 45736. https://doi.org/10.1002/app.45736

[25] Z. Dai, X. Shi, H. Liu, H. Li, Y. Han, J. Zhou, High-strength lignin-based carbon

fibers *via* a low-energy method, RSC Adv. 8 (2018) 1218–1224.
https://doi.org/10.1039/C7RA10821D

[26] M. Nar, H.R. Rizvi, R.A. Dixon, F. Chen, A. Kovalcik, N. D'Souza, Superior plant
based carbon fibers from electrospun poly-(caffeyl alcohol) lignin, Carbon 103
(2016) 372–383. https://doi.org/10.1016/j.carbon.2016.02.053

[27] W.-J. Liu, H. Jiang, H.Q. Yu, Thermochemical conversion of lignin to functional
materials: a review and future directions, Green Chem. 17 (2015) 4888–4907.
https://doi.org/10.1039/C5GC01054C

[28] S. Hu, S. Zhang, N. Pan, Y.-L. Hsieh, High energy density supercapacitors from
lignin derived submicron activated carbon fibers in aqueous electrolytes, J. Power
Sources 270 (2014) 106–112. https://doi.org/10.1016/j.jpowsour.2014.07.063

[29] M. Ago, M. Borghei, J.S. Haataja, O.J. Rojas, Mesoporous carbon soft-templated
from lignin nanofiber networks: microphase separation boosts supercapacitance in
conductive electrodes, RSC Adv. 6 (2016) 85802–85810.
https://doi.org/10.1039/C6RA17536H

[30] R. Ruiz-Rosas, M.J. Valero-Romero, D. Salinas-Torres, J. Rodríguez-Mirasol, T.
Cordero, E. Morallón, D. Cazorla-Amorós, Electrochemical performance of
hierarchical porous carbon materials obtained from the infiltration of lignin into
zeolite templates, ChemSusChem 7 (2014) 1458–1467.
https://doi.org/10.1002/cssc.201301408

[31] J. Tian, Z. Liu, Z. Li, W. Wang, H. Zhang, Hierarchical S-doped porous carbon
derived from by-product lignin for high-performance supercapacitors, RSC Adv. 7
(2017) 12089–12097. https://doi.org/10.1039/C7RA00767A

[32] A.M. Navarro-Suárez, J. Carretero-González, V. Roddatis, E. Goikolea, J. Ségalini,
E. Redondo, T. Rojo, R. Mysyk, Nanoporous carbons from natural lignin: Study of
structural–textural properties and application to organic-based supercapacitors,
RSC Adv. 4 (2014) 48336–48343. https://doi.org/10.1039/C4RA08218D

[33] D. Saha, Y. Li, Z. Bi, J. Chen, J.K. Keum, D.K. Hensley, H.A. Grappe, H.M.
Meyer, S. Dai, M.P. Paranthaman, A.K. Naskar, Studies on supercapacitor
electrode material from activated lignin-derived mesoporous carbon, Langmuir 30
(2014) 900–910. https://doi.org/10.1021/la404112m

[34] I. Isaev, G. Salitra, A. Soffer, Y.S. Cohen, D. Aurbach, J. Fischer, A new approach
for the preparation of anodes for Li-ion batteries based on activated hard carbon

cloth with pore design, J. Power Sources 119-121 (2003) 28-33.
https://doi.org/10.1016/S0378-7753(03)00119-8

[35] G.T.K. Fey, Y.D. Cho, C.L. Chen, K.P. Huang, Y.C. Lin, T.P. Kumar, S.H. Chan, Pyrolytic carbons from porogen-treated rice husk as lithium-insertion anode materials, Int. J. Chem. Eng. Appl. (2011) 20–25.
https://doi.org/10.7763/IJCEA.2011.V2.69

[36] Z.W. He, Q.F. Lü, Q. Lin, Fabrication, characterization and application of nitrogen-containing carbon nanospheres obtained by pyrolysis of lignosulfonate/poly(2-ethylaniline), Bioresour. Technol. 127 (2013) 66–71.
https://doi.org/10.1016/j.biortech.2012.09.132

[37] S.X. Wang, L. Yang, L.P. Stubbs, X. Li, C. He, Lignin-derived fused electrospun carbon fibrous mats as high performance anode materials for lithium ion batteries, ACS Appl. Mater. Interfaces 5 (2013) 12275–12282.
https://doi.org/10.1021/am4043867

[38] F. Yu, Y. Li, M. Jia, T. Nan, H. Zhang, S. Zhao, Q. Shen, Elaborate construction and electrochemical properties of lignin-derived macro-/micro-porous carbon-sulfur composites for rechargeable lithium-sulfur batteries: The effect of sulfur-loading time, J. Alloys Compd. 709 (2017) 677–685.
https://doi.org/10.1016/j.jallcom.2017.03.204

[39] H. Li, D. Yuan, C. Tang, S. Wang, J. Sun, Z. Li, T. Tang, F. Wang, H. Gong, C. He, Lignin-derived interconnected hierarchical porous carbon monolith with large areal/volumetric capacitances for supercapacitor, Carbon 100 (2016) 151–157.
https://doi.org/10.1016/j.carbon.2015.12.075

[40] F.J. García-Mateos, R. Berenguer, M.J. Valero-Romero, J. Rodríguez-Mirasol, T. Cordero, Phosphorus functionalization for the rapid preparation of highly nanoporous submicron-diameter carbon fibers by electrospinning of lignin solutions, J. Mater. Chem. A. 6 (2018) 1219–1233.
https://doi.org/10.1039/C7TA08788H

[41] R. Berenguer, R. Ruiz-Rosas, A. Gallardo, D. Cazorla-Amorós, E. Morallón, H. Nishihara, T. Kyotani, J. Rodríguez-Mirasol, T. Cordero, Enhanced electro-oxidation resistance of carbon electrodes induced by phosphorus surface groups, Carbon 95 (2015) 681–689. https://doi.org/10.1016/j.carbon.2015.08.101

[42] C. Huang, A.M. Puziy, O.I. Poddubnaya, D. Hulicova-Jurcakova, M. Sobiesiak, B.

Gawdzik, Phosphorus, nitrogen and oxygen co-doped polymer-based core-shell carbon sphere for high-performance hybrid supercapacitors, Electrochim. Acta 270 (2018) 339–351. https://doi.org/10.1016/j.electacta.2018.02.115

[43] C. Lai, Z. Zhou, L. Zhang, X. Wang, Q. Zhou, Y. Zhao, Y. Wang, X.F. Wu, Z. Zhu, H. Fong, Free-standing and mechanically flexible mats consisting of electrospun carbon nanofibers made from a natural product of alkali lignin as binder-free electrodes for high-performance supercapacitors, J. Power Sources 247 (2014) 134–141. https://doi.org/10.1016/j.jpowsour.2013.08.082

[44] W. Liu, Y. Yao, O. Fu, S. Jiang, Y. Fang, Y. Wei, X. Lu, Lignin-derived carbon nanosheets for high-capacitance supercapacitors, RSC Adv. 7 (2017) 48537–48543. https://doi.org/10.1039/C7RA08531A

[45] C. Xiong, W. Zhong, Y. Zou, J. Luo, W. Yang, Electroactive biopolymer/graphene hydrogels prepared for high-performance supercapacitor electrodes, Electrochim. Acta 211 (2016) 941–949. https://doi.org/10.1016/j.electacta.2016.06.117

[46] G. Milczarek, O. Inganäs, Renewable cathode materials from biopolymer/conjugated polymer interpenetrating networks., Science 335 (2012) 1468–71. https://doi.org/10.1126/science.1215159

[47] H. Xu, H. Jiang, X. Li, G. Wang, Synthesis and electrochemical capacitance performance of polyaniline doped with lignosulfonate, RSC Adv. 5 (2015) 76116–76121. https://doi.org/10.1039/C5RA12292A

[48] P. Saini (Eds.), Fundamentals of conjugated polymer blends, copolymers and composites : synthesis, properties and applications, Wiley, 2015. https://doi.org/10.1002/9781119137160

[49] L. Zhu, L. Wu, Y. Sun, M. Li, J. Xu, Z. Bai, G. Liang, L. Liu, D. Fang, W. Xu, Cotton fabrics coated with lignosulfonate-doped polypyrrole for flexible supercapacitor electrodes, RSC Adv. 4 (2014) 6261. https://doi.org/10.1039/c3ra47224h

[50] T.T. Lin, W.D. Wang, Q.F. Lü, H.B. Zhao, X. Zhang, Q. Lin, Graphene-wrapped nitrogen-containing carbon spheres for electrochemical supercapacitor application, J. Anal. Appl. Pyrolysis. 113 (2015) 545–550. https://doi.org/10.1016/j.jaap.2015.03.013

[51] J. Jin, B. Yu, Z. Shi, C. Wang, C. Chong, Lignin-based electrospun carbon nanofibrous webs as free-standing and binder-free electrodes for sodium ion

batteries, J. Power Sources 272 (2014) 800–807.
https://doi.org/10.1016/j.jpowsour.2014.08.119

[52] J.M. Rosas, R. Berenguer, M.J. Valero-Romero, J. RodrÃ-guez-Mirasol, T.
      Cordero, Preparation of different carbon materials by thermochemical conversion
      of lignin, Front. Mater. 1 (2014) 29. https://doi.org/10.3389/fmats.2014.00029

[53] H. Lu, A. Cornell, F. Alvarado, M. Behm, S. Leijonmarck, J. Li, P. Tomani, G.
      Lindbergh, H. Lu, A. Cornell, F. Alvarado, M. Behm, S. Leijonmarck, J. Li, P.
      Tomani, G. Lindbergh, Lignin as a binder material for eco-friendly li-ion batteries,
      Materials 9 (2016) 127. https://doi.org/10.3390/ma9030127

[54] T. Chen, Q. Zhang, J. Pan, J. Xu, Y. Liu, M. Al-Shroofy, Y.T. Cheng, Low-
      temperature treated lignin as both binder and conductive additive for silicon
      nanoparticle composite electrodes in lithium-ion batteries, ACS Appl. Mater.
      Interfaces 8 (2016) 32341–32348. https://doi.org/10.1021/acsami.6b11500

[55] A.M. Navarro-Suárez, J. Carretero-González, T. Rojo, M. Armand, Poly(quinone-
      amine)/nanocarbon composite electrodes with enhanced proton storage capacity, J.
      Mater. Chem. A. 5 (2017) 23292–23298. https://doi.org/10.1039/C7TA08489G

[56] K. Wang, M. Xu, Y. Gu, Z. Gu, Q.H. Fan, Symmetric supercapacitors using urea-
      modified lignin derived N-doped porous carbon as electrode materials in liquid and
      solid electrolytes, J. Power Sources 332 (2016) 180–186.
      https://doi.org/10.1016/j.jpowsour.2016.09.115

[57] X. Xu, J. Zhou, D.H. Nagaraju, L. Jiang, V.R. Marinov, G. Lubineau, H.N.
      Alshareef, M. Oh, Flexible, highly graphitized carbon aerogels based on bacterial
      cellulose/lignin: Catalyst-free synthesis and its application in energy storage
      devices, Adv. Funct. Mater. 25 (2015) 3193–3202.
      https://doi.org/10.1002/adfm.201500538

[58] S. Leguizamon, K.P. Díaz-Orellana, J. Velez, M.C. Thies, M.E. Roberts, High
      charge-capacity polymer electrodes comprising alkali lignin from the kraft process,
      J. Mater. Chem. A. 3 (2015) 11330–11339. https://doi.org/10.1039/C5TA00481K

[59] S. Admassie, A. Elfwing, E.W.H. Jager, Q. Bao, O. Inganäs, A renewable
      biopolymer cathode with multivalent metal ions for enhanced charge storage, J.
      Mater. Chem. A. 2 (2014) 1974–1979. https://doi.org/10.1039/C3TA13876C

[60] M. Klose, R. Reinhold, F. Logsch, F. Wolke, J. Linnemann, U. Stoeck, S. Oswald,
      M. Uhlemann, J. Balach, J. Markowski, P. Ay, L. Giebeler, Softwood Lignin as a

sustainable feedstock for porous carbons as active material for supercapacitors using an ionic liquid electrolyte, ACS Sustain. Chem. Eng. 5 (2017) 4094–4102. https://doi.org/10.1021/acssuschemeng.7b00058

[61] W. Zhang, M. Zhao, R. Liu, X. Wang, H. Lin, Hierarchical porous carbon derived from lignin for high performance supercapacitor, Colloids Surfaces A Physicochem. Eng. Asp. 484 (2015) 518–527. https://doi.org/10.1016/j.colsurfa.2015.08.030

[62] D. Salinas-Torres, R. Ruiz-Rosas, M.J. Valero-Romero, J. Rodríguez-Mirasol, T. Cordero, E. Morallón, D. Cazorla-Amorós, Asymmetric capacitors using lignin-based hierarchical porous carbons, J. Power Sources 326 (2016) 641–651. https://doi.org/10.1016/j.jpowsour.2016.03.096

Biomass Based Energy Storage Materials
Materials Research Foundations **78** (2020) 111-123

Materials Research Forum LLC
https://doi.org/10.21741/9781644900871-5

Chapter 5

Bamboo Derived Materials for Energy Storage

Sivagaami Sundari Gunasekaran[1], Thileep Kumar Kumaresan[1], Shanmugaraj Andikaddu Masilamanai[1], Kalaivani Raman[1], Raghu Subash Chandra Bose[1]*

[1] Department of Chemistry, Vels Institute of Science, Technology and Advanced Studies (VISTAS), Chennai, 600117, India

*subraghu_0612@yahoo.co.in

Abstract

Natural bamboo is an eco-friendly, widely distributed and multifunctional plant which has fast growth rate, short maturation cycle and high production yield. Recycling the bamboo wastes can make the process cost-effective and environment friendly. Added, the bamboo waste can be carbonized yielding low-cost carbon material which can be employed as electrode materials for energy storage devices like supercapacitors, batteries and fuel cells. Bamboo-based materials have been chosen for the electrode fabrication by virtue of its unique fibrous structure and due to the larger inner surface area provided by its tubular architecture. Compared to reported activated carbon, carbons derived from bamboo wastes have shown great promises for energy related applications. In this chapter, the physico-chemical characteristics of activated carbons derived from bamboo wastes for the supercapacitor application is discussed.

Keywords

Activated Carbon, Bamboo Stick, Supercapacitor, Specific Capacitance, Electrode

Contents

1. Introduction 112

2. Fabrication of electrode material for supercapacitor application 113

3. Physical characterization 114

4. Electrochemical measurements 116

Conclusion 118

References **119**

1. Introduction

Supercapacitors (or) ultracapacitors are catching attention in electronic, electrical and automobile sectors due to their superior inheritance of power density, cycling stability, rapid charging-discharging and profound maintenances. Based upon choosing of the electrode materials, the performance of the energy storage materials is reflected [1]. In order to enhance the specific capacitance, energy and power density of supercapacitor, the electrode material should be precisely synthesized and accurately fabricated.

Bamboo waste is chosen for the preparation of electrode material due to their abundance and renewable property. This biomass can be carbonized even at very high temperature to produce carbon or graphene structured nanoporous carbonaceous material for energy related applications, especially supercapacitors. The bamboo derived activated carbon has unique structural properties like interconnectivity, multi-channels and high porous nature. The porosity activates the electrochemical behavior of the bamboo precursor carbon that can promote the fast access to electrolyte and ion-diffusion [2-6].

In this chapter, we will summarize the fabrication method to synthesize activated carbon from *bamboo waste*. The pore formation and increase in electrochemical performance are discussed briefly. This chapter focuses on the chemical activation of bamboo waste derived carbon (char) with the aid of activating agents like Potassium hydroxide (KOH), potassium ferrocyanide ($K_3[Fe(CN)_6]$) and zinc chloride ($ZnCl_2$) [7-8].

The recent progress and applications in *bamboo-derived carbon* for energy storage systems are discussed here. The chapter finally concludes the perspective of the bamboo-derived carbon for *supercapacitor applications*. The different activation procedures available in the literatures are illustrated in the flow chart *(Figure 1)* which deals with all activating agents, both physical and chemical treatments, even the combined activation methods.

The bamboo derived carbons have been applied in energy storage devices in recent years [9-11]. Jian Jiang et al. [13] have fabricated highly efficient carbon fiber electrode from *bamboo biomass* (chopsticks) for *lithium-ion battery*. Solid-state flexible supercapacitor developed from bamboo biomass by Yongming Sun et al, showed better performance [7]. Camila Zequine et al. fabricated flexible supercapacitor using potassium hydroxide (KOH) activated carbon material derived from bamboo which exhibited improved electrochemical capacitive performance [8]. Yuxiang Wen et al. [14] and G.Y. Ping et.al,

Biomass Based Energy Storage Materials   Materials Research Forum LLC
Materials Research Foundations **78** (2020) 111-123   https://doi.org/10.21741/9781644900871-5

[15] reported new strategies for converting bamboo waste into carbon fiber for the fabrication of supercapacitor.

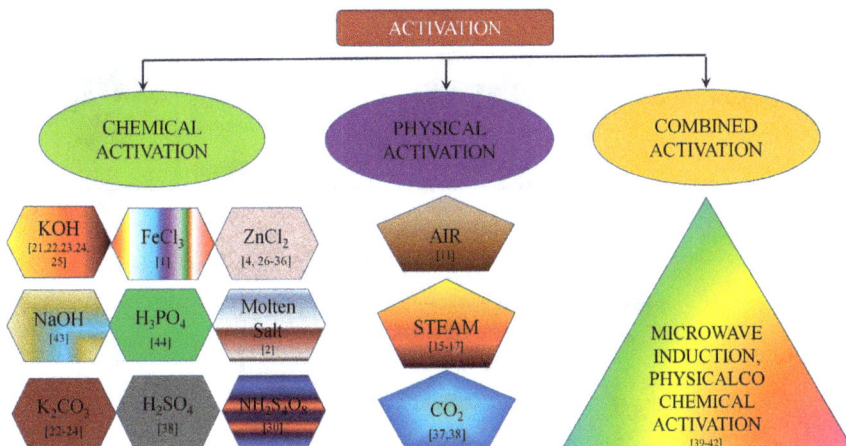

Figure 1. Schematic Representation of Different Activation Treatments of Bamboo-Derived Activated carbon.

2.   Fabrication of electrode material for supercapacitor application

Motivated by these provocative studies with fascinating results, we employed two different methods to prepare the activated carbon as follows (a) chemical activating and (b) combined hydrothermal with chemical activation for the fabrication of electrode materials for supercapacitor applications.

The bamboo wastes are carbonized by following simple pyrolysis method keeping the waste in a muffle furnace at *$300^{o}C$ for almost 3 hours [16-18]*. The temperature was optimized by carrying out different slots of experiments. The schematic representation for carbonization of bamboo waste is shown in *Figure 2*.

The char produced after the pyrolysis of the bamboo biomass is then subjected for activation which is done using the vertical tubular furnace. Different activation temperatures can be employed for the chemical activation procedures. Also, various inert gases like carbon-di-oxide, argon, nitrogen can be employed for activation purpose [18-20]. Herein, the bamboo derived char *(shown in Figure 2)*, is subjected to activation with

Biomass Based Energy Storage Materials   Materials Research Forum LLC
Materials Research Foundations **78** (2020) 111-123   https://doi.org/10.21741/9781644900871-5

three chemical activating agents such potassium hydroxide (KOH), potassium ferrocyanide ($K_3[Fe(CN)_6]$) and zinc chloride ($ZnCl_2$) [21-28].

Figure 2. Schematic Representation of Carbonization and Activation of the Bamboo-derived activated carbon

3. Physical characterization

The X-ray diffraction study is employed for identifying the crystal nature and property of the material. The X-ray diffraction studies for the KOH activated carbon from bamboo biomass revealed the peaks at $22.3°$ and $43°$. The XRD spectrum is shown in *Figure 3(a)*. The X-ray spectrum of $KOH+K_3[Fe(CN)_6]$ catalyzed activated carbon showed two peaks at $25.8°$ and $43°$ which represents the partially graphitic nature of carbon. The potassium ferrocyanide is converted to iron-oxide and plays a role of electrocatalyst. The zinc-catalyzed activated carbon showed two eminent peaks at $23.8°$ and $45°$. It is clear from these results that, when the activated carbon derived from *bamboo biomass* treated with iron, it becomes mixture of amorphous and crystalline property, while treating with zinc salt, it becomes highly crystalline in nature [29-32].

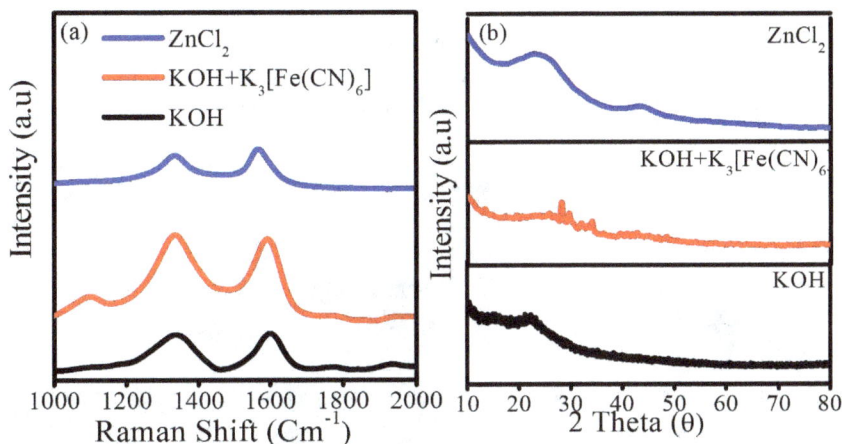

Figure 3. (a) X-ray Diffraction spectrum and (b) Raman Spectrum of the prepared bamboo-derived activated carbonss

Generally, the Raman spectroscopy identifies the degree of graphitization and degree of disorder which can identify the nature of the prepared materials. D band represents the degree of disorder peak and G band represents the degree of graphitic peak. All the three prepared activated carbons (KOH, KOH+K_3[Fe (CN)$_6$], and ZnCl$_2$) shows eminent peaks of D band and G band depicted in *Figure 3(b)*. Compared to KOH and KOH+K_3[Fe (CN)$_6$ catalyzed carbons, the graphitic nature is boosted in the zinc catalyzed carbon. Both the G and D bands are almost equal in the KOH and KOH+K_3[Fe (CN)$_6$ catalyzed activated carbons derived from bamboo waste. The presence of activating agents influences the graphitic nature of the material which almost changed its crystal nature of the materials. [33-34]

The surface morphologies of the prepared activated carbons in presence of KOH, KOH+K_3[Fe (CN)$_6$] and ZnCl$_2$ are shown *in Figure 4(a-c)*. It is seen that, as the activating agents are differed, the surface morphology is getting different. Different structures are formed. Added, the pores are also formed clearly. When activating with potassium hydroxide, foam or paper-like morphology is obtained and the pores are not formed clearly. In KOH+K_3[Fe (CN)$_6$] and ZnCl$_2$ activated carbon, the formations of pores are apparent. That is, the hallow morphology is distinctly seen. Added, the micro-pores are only formed. In the case of ZnCl$_2$-catalyzed activated carbon prepared from bamboo waste, flake like surface morphology is obtained. Also, there is a uniformity

Biomass Based Energy Storage Materials
Materials Research Foundations **78** (2020) 111-123

Materials Research Forum LLC
https://doi.org/10.21741/9781644900871-5

trend constituted in the surface of the carbon. Concluding the physical characterization and surface morphology analysis of the prepared activated carbon from bamboo waste, it is clearly confirmed that, these changes and developments in the carbon would explicitly compliment and reinforce the electrochemical properties and energy storage (supercapacitor) performances.

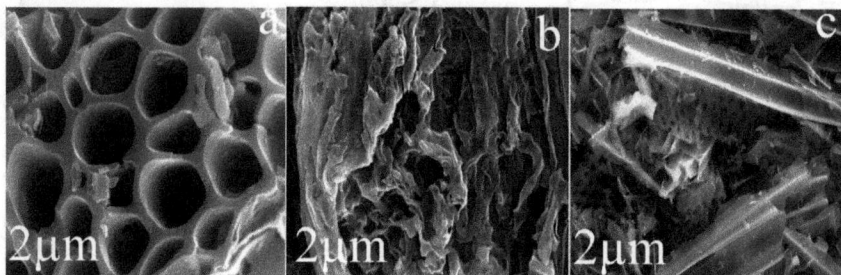

Figure 4. Scanning Electron Microscopy images of (a) KOH activated (b) KOH+K₃[Fe(CN)₆]- activated and (c) Zncl₂-activated prepared bamboo-derived carbon

4. Electrochemical measurements

The cyclic voltammetry for the three different prepared activated carbons derived from the bamboo biomass is depicted in *Figure 5(a)*. Two-electrode configuration with the aid of coin cell CR2032 was made within the voltage range of 0 to 1V in the presence of aqueous electrolyte (6M KOH) [35-39]. At the scan rate of $50mVs^{-1}$, $ZnCl_2$-catalyzed carbon has the perfect rectangular shape (symmetry) with little negligible infra-red (IR) drop. It means that, the resistance in the $ZnCl_2$-catalyzed material is exorbitantly very low compared to other two activated carbons. This implies that the conductivity, ie. the movement of ions is very high for zinc-catalyzed carbon and acutely low for the KOH-treated and KOH+K₃[Fe (CN)₆] treated carbons. The specific capacitance values were approximately, 120, 170 and 180 Fg^{-1} for KOH, KOH+K₃[Fe (CN)₆] and zinc-catalyzed activated carbons, respectively.

To evaluate the electrochemical capacitance of prepared materials under controlled current condition, the galvanostatic charge- discharge studies were employed [40]. Generally, when the discharge takes longer time, it implies that the device can operate for a longer period of time with better and improved performance [41-43]. The results shown in *Figure 5(b)* reveal that the $ZnCl_2$ activated carbon has the maximum seconds for discharging time of approximately 120 seconds, compared to the other activated carbon (KOH and KOH+K₃[Fe (CN)₆]) resulting its discharging capacity around 80 and 50 seconds, respectively. The KOH activated carbon outputs the energy density of 9.7 $Whkg^{-1}$ at $1Ag^{-1}$ and power density of 1.9 $kWkg^{-1}$ at 10 Ag^{-1}. An energy density of 12.7 $Whkg^{-1}$ at $1Ag^{-1}$ power density of 2.1 $kWkg^{-1}$ was calculated for the iron-activated carbon. Higher values of energy and power densities were calculated for the $ZnCl_2$ activated carbon derived from bamboo biomass. Higher energy and power density obtained were 12.8 $Whkg^{-1}$ at $1Ag^{-1}$ and 2.7 $kWkg^{-1}$, respectively.

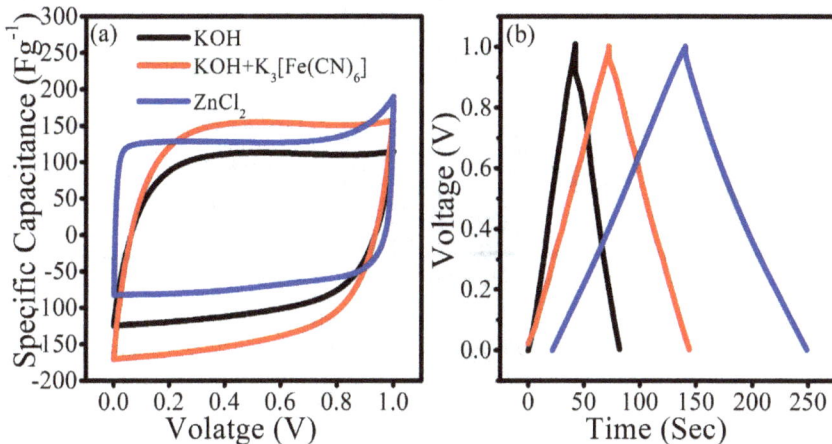

Figure 5. Schematic Representation of (a) cyclic voltammetry curve and (b) galvanostatic charge-discharge curve of prepared bamboo-derived activated carbons

Nyquist plot represents the vector response of the output system showing the relationship between the feedback and the gain. The semi-circle in the plot formed only for the ideal case of the material which is suitable for the electrochemical part. Here, generally it is concluded that, smaller the semi-circle is, the lower is the resistance in the material and thus higher is the conductivity of the material and the material is ideal for supercapacitor applications. The elongated semi-circles in the plot represent the effect of the pore formation in the material. The electrochemical impedance spectra of the prepared activated carbon materials are represented in the *Figure 6(a)*. The impedances of 35, 5.5 and 4 Ω are produced for the KOH, iron and zinc activated carbons from bamboo biomass.

The KOH activated carbon showed 89% cycling stability retention for about 10,000 cycles and the KOH+K_3[Fe (CN)$_6$] catalyzed carbon has the retention of 95 % and the higher stability retention was got for $ZnCl_2$-catalyzed carbon material synthesized from the bamboo biomass of about 98% which is clearly shown in *Figure 6(b)*.

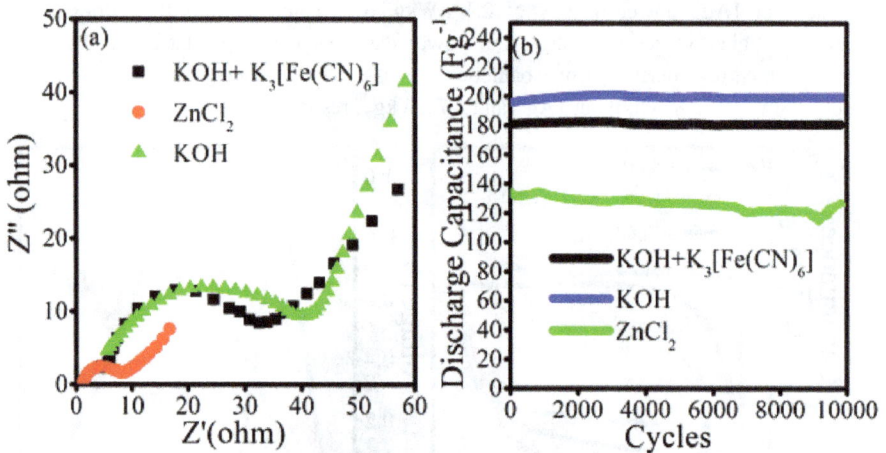

Figure 6. Schematic Representation of (a) Electrochemical Impedance spectrum and (b) cycling study of the prepared bamboo-derived activated carbons

Conclusion

Many novel and efficient electrode materials for energy storage systems can be obtained from bamboo biomass, which is the bioactive substance. The recycling of the bamboo waste can be a green value-added auxiliary implementation window for the biomass-based products to convert it as an advanced and significant activated carbon as high-performance energy storage electrode materials, specifically for supercapacitor electrode materials. Different electrode materials of activated carbons derived from biomass have been reviewed. The enhancement in the electrochemical performance of the chemically activated carbons has been explained clearly. Additionally, the formation of the pores is explained in detailed with the surface morphological study by the aid of scanning electron microscopy technique. In addition, the electrochemical performance of the prepared activated carbons is summarized. From these results, it can be evident that the selection of chemical reagents or activating agents, activating temperature, after-treatment processes, the electrochemical performance of the energy storage can be improved substantially. From the surface morphological study, it can be concluded that there is a homogeneous pore formation in the internal sites of the material. Thus, this piece of work with combined hydrothermal and chemical activation can be considered as an efficient and green way for the production of bamboo derived carbon and improved electrochemical performance of the energy storage device.

References

[1]   Gunasekaran, S. Sundari, S.K. Elumalali, T.K. Kumaresan, R. Meganathan, A. Ashok, V. Pawar, K. Vediappan, Partially graphitic nanoporous activated carbon prepared from biomass for supercapacitor application, Mater. Lett. 218 (2018) 165-168. https://doi.org/10.1016/j.matlet.2018.01.172

[2]   Kumar, K. Thileep, G.S. Sundari, E.S. Kumar, A. Ashwini, M. Ramya, P. Varsha, R. Kalaivani, Synthesis of nanoporous carbon with new activating agent for high-performance supercapacitor, Mater. Lett. 218 (2018) 181-184. https://doi.org/10.1016/j.matlet.2018.02.017

[3]   E.V. Senthilkumar, B. Sivasankar, R. Kohakade, K. Thileepkumar, M. Ramya, G.S. Sundari, S. Raghu, R.A. Kalaivani, Synthesis of nanoporous graphene and their electrochemical performance in a symmetric supercapacitor, Appl. Surface Sci. 460 (2018) 17-24. https://doi.org/10.1016/j.apsusc.2017.10.221

[4]   S.S. Gunasekaran, R.S. Bose, K. Raman, Electrochemical capacitive performance of zncl$_2$ activated carbon derived from bamboo bagasse in aqueous and organic electrolyte, Orient J. Chem. 35 (2019) 350136. https://doi.org/10.13005/ojc/350136

[5]   Tian, Weiqian, Q. Gao, Y. Tan, K. Yang, L. Zhu, C. Yang, H. Zhang, Bio-inspired beehive-like hierarchical nanoporous carbon derived from bamboo-based industrial by-product as a high performance supercapacitor electrode material, J. Mater. Chem. A 3 (2015) 5656-5664. https://doi.org/10.1039/C4TA06620K

[6]   Tang, Wangjia, Y. Zhang, Y. Zhong, T. Shen, X. Wang, X. Xia, J. Tu, Natural biomass-derived carbons for electrochemical energy storage, Mater. Res. Bull. 88 (2017) 234-241. https://doi.org/10.1016/j.materresbull.2016.12.025

[7]   S. Yongming, R.B. Sills, X. Hu, Z.W. Seh, X. Xiao, H. Xu, W. Luo, A bamboo-inspired nanostructure design for flexible, foldable, and twistable energy storage devices, Nano Lett. 15 (2015) 3899-3906. https://doi.org/10.1021/acs.nanolett.5b00738

[8]   Z. Camila, C.K. Ranaweera, Z. Wang, S. Singh, P. Tripathi, O.N. Srivastava, B.K. Gupta, High per formance and flexible supercapacitors based on carbonized bamboo fibers for wide temperature applications, Sci. Rep. 6 (2016) 31704. https://doi.org/10.1038/srep31704

[9]   W. Yuxiang, T. Qin, Z. Wang, X. Jiang, S. Peng, J. Zhang, J. Hou, F. Huang, D. He, G.Cao, Self-supported binder-free carbon fibers/MnO₂electrodes derived from disposable bamboo chopsticks for high-performance supercapacitors, J. Alloys Compd. 699 (2017) 126-135. https://doi.org/10.1016/j.jallcom.2016.12.330

[10]   C. Hao, D. Liu, Z. Shen, B. Bao, S. Zhao, L. Wu, Functional biomass carbons with hierarchical porous structure for supercapacitor electrode materials, Electrochim. Acta 180 (2015) 241-251. https://doi.org/10.1016/j.electacta.2015.08.133

[11]   L. Yuanyuan, L. Wang, B. Gao, X.Li, Q. Cai, Q. Li, X. Peng, K. Huo, P.K. Chu, Hierarchical porous carbon materials derived from self-template bamboo leaves for lithium–sulfur batteries, Electrochim. Acta 229 (2017) 352-360. https://doi.org/10.1016/j.electacta.2017.01.166

[12]   H. Wei, S. Deng, B. Hu, Z. Chen, B. Wang, J. Huang, G. Yu, Granular bamboo-derived activated carbon for high CO₂ adsorption: the dominant role of narrow micropores, ChemSusChem 12 (2012) 2354-2360. https://doi.org/10.1002/cssc.201200570

[13]   J. Jiang, J. Zhu, W. Ai, Z. Fan, X. Shen, C. Zou, J. Liu, H. Zhang, T. Yu, Evolution of disposable bamboo chopsticks into uniform carbon fibers: a smart strategy to fabricate sustainable anodes for Li-ion batteries, Energy Environ. Sci. 8 (2014) 2670-2679. https://doi.org/10.1039/C4EE00602J

[14]   Z. Guoxiang, H. Chen, W. Liu, D. Wang, Y. Wang, Bamboo chopsticks-derived porous carbon microtubes/flakes composites for supercapacitor electrodes, Mater. Lett. 185 (2016) 359-362. https://doi.org/10.1016/j.matlet.2016.09.045

[15]   G.Y. Ping, Z.B. Zhai, K.J. Huang, Y.Y. Zhang, Energy storage applications of biomass-derived carbon materials: Batteries and supercapacitors, New J. Chem. 41 (2017) 11456-11470. https://doi.org/10.1039/C7NJ02580G

[16]   Y. Yinglin, M. Shi, Y. Wei, C. Zhao, M. Carnie, R. Yang, Y. Xu, Process optimization for producing hierarchical porous bamboo-derived carbon materials with ultrahigh specific surface area for lithium-sulfur batteries, J. Alloys Compd. 738 (2018) 16-24. **https://doi.org/10.1016/j.jallcom.2017.11.212**

[17]   L.Q. Chao, T. Liu, D.P. Liu, Z.J. Li, X.B. Zhang, Y. Zhang, A flexible and wearable lithium–oxygen battery with record energy density achieved by the interlaced architecture inspired by bamboo slips, Adv. Mater. 28 (2016) 8413-8418. https://doi.org/10.1002/adma.201602800

[18]   C. Xiufang, J. Zhang, B. Zhang, S. Dong, X. Guo, X. Mu, B. Fei, A novel
        hierarchical porous nitrogen-doped carbon derived from bamboo shoot for high
        performance supercapacitor, Sci. Rep. 7 (2017) 7362.
        https://doi.org/10.1038/s41598-017-06730-x

[19]   C. Dengyu, D. Liu, H. Zhang, Y. Chen, Q. Li, Bamboo pyrolysis using TG–FTIR
        and a lab-scale reactor: Analysis of pyrolysis behavior, product properties, and
        carbon and energy yields, Fuel 148 (2015) 79-86.
        https://doi.org/10.1016/j.fuel.2015.01.092

[20]   S. Daniel, M. Escala, K. Supawittayayothin, N. Tippayawong, Characterization of
        biochar from hydrothermal carbonization of bamboo, Int. J. Energy Environ. 2
        (2011): 647-652.

[21]   W. Huanlei, Q. Gao, J. Hu, High hydrogen storage capacity of porous carbons
        prepared by using activated carbon, J. Am. Chem. Soc.131 (2009) 7016-7022.
        https://doi.org/10.1021/ja8083225

[22]   O. Toshiro, R. Tanibata, M. Itoh, Production and adsorption characteristics of
        MAXSORB: high-surface-area active carbon, Gas Sep. Purif. 7 (1993) 241-245.
        https://doi.org/10.1016/0950-4214(93)80024-Q

[23]   R. Pinero, E.P. Azais, T. Cacciaguerra, D. Cazorla-Amorós, A. Linares-Solano, F.
        Béguin, KOH and NaOH activation mechanisms of multiwalled carbon nanotubes
        with different structural organisation, Carbon 43 (2005) 786-795.
        https://doi.org/10.1016/j.carbon.2004.11.005

[24]   Q. Wenming, S.H. Yoon, I. Mochida, KOH activation of needle coke to develop
        activated carbons for high-performance EDLC, Energy Fuels 20 (2006) 1680-
        1684. https://doi.org/10.1021/ef0503131

[25]   L. Castello, J.M. Calo, D. Cazorla-Amoros, A. Linares-Solano, Carbon activation
        with KOH as explored by temperature programmed techniques, and the effects of
        hydrogen, Carbon 45 (2007) 2529-2536.
        https://doi.org/10.1016/j.carbon.2007.08.021

[26]   V. Subramanian, C. Luo, A.M. Stephan, K.S. Nahm, S. Thomas, B. Wei,
        Supercapacitors from activated carbon derived from banana fibers, J. Phy. Chem.
        C 111 (2007) 7527-7531. https://doi.org/10.1021/jp067009t

[27]   C. Lulu, P. Guo, R. Wang, L. Ming, F. Leng, H. Li, X.S. Zhao, Electrocapacitive
        properties of supercapacitors based on hierarchical porous carbons from chestnut

shell, Colloids Surf. A Physicochem. Eng. Asp. 446 (2014) 127-133.
https://doi.org/10.1016/j.colsurfa.2014.01.057

[28] Z. Dengyun, H. Du, B. Li, Y. Zhu, F. Kang, Porous graphitic carbons prepared by combining chemical activation with catalytic graphitization, Carbon 49 (2011) 725-729. https://doi.org/10.1016/j.carbon.2010.09.057

[29] H. Jianhua, C. Cao, F. Idrees, X. Ma, Hierarchical porous nitrogen-doped carbon nanosheets derived from silk for ultrahigh-capacity battery anodes and supercapacitors, ACS Nano 9 (2015) 2556-2564.
https://doi.org/10.1021/nn506394r

[30] S. Li, C. Tian, M. Li, X. Meng, L. Wang, R. Wang, J. Yin, H. Fu, From coconut shell to porous graphene-like nanosheets for high-power supercapacitors, J. Mater. Chem.A 21 (2013) 6462-6470. https://doi.org/10.1039/c3ta10897j

[31] R.E. Thomas, D. Hulicova-Jurcakova, Erika Fiset, Zhonghua Zhu,Gao Qing Lu, Double-layer capacitance of waste coffee ground activated carbons in an organic electrolyte, Electrochem. Commun. 11 (2009) 974-977.
https://doi.org/10.1016/j.elecom.2009.02.038

[32] T.E. Rufford, D.H. Jurcakova, Z. Zhu, G.Q. Lu, Nanoporous carbon electrode from waste coffee beans for high performance supercapacitors, Electrochem. Commun. 10 (2008) 1594-1597. https://doi.org/10.1016/j.elecom.2008.08.022

[33] T.E. Rufford, D.H. Jurcakova, K. Khosla, Z. Zhu, G.Q. Lu, Microstructure and electrochemical double-layer capacitance of carbon electrodes prepared by zinc chloride activation of sugar cane bagasse, J. Power Sources 195 (2010) 912-918.
https://doi.org/10.1016/j.jpowsour.2009.08.048

[34] L. Zichao, K. Zhai, G. Wang, Q. Li, P. Guo, Preparation and electrocapacitive properties of hierarchical porous carbons based on loofah sponge, Materials 9 (2016) 912. https://doi.org/10.3390/ma9110912

[35] W. Xianjun, X. Jiang, J. Wei, S. Gao, Functional groups and pore size distribution do matter to hierarchically porous carbons as high-rate-performance supercapacitors, Chem. Mater. 28 (2016) 445-458.
https://doi.org/10.1021/acs.chemmater.5b02336

[36] C. Haiqun, M.B. Müller, K.J. Gilmore, G.G. Wallace, D. Li, Mechanically strong, electrically conductive, and biocompatible graphene paper, Adv. Mater. 20 (2008) 3557-3561. https://doi.org/10.1002/adma.200800757

[37]   W. Huanlei, Z. Li, D. Mitlin, Tailoring biomass-derived carbon nanoarchitectures
        for high-performance supercapacitors, ChemElectroChem 1 (2014) 332-337.
        https://doi.org/10.1002/celc.201300127

[38]   L. Zhi, L. Zhang, B.S. Amirkhiz, X. Tan, Z. Xu, H. Wang, B.C. Olsen, C.M.B.
        Holt, D. Mitlin, Carbonized chicken eggshell membranes with 3D architectures as
        high-performance electrode materials for supercapacitors, Adv. Energy Mater. 2
        (2012) 431-437. https://doi.org/10.1002/aenm.201100548

[39]   S. Marta, A.B. Fuertes, The production of carbon materials by hydrothermal
        carbonization of cellulose, Carbon 47 (2009) 2281-2289.
        https://doi.org/10.1016/j.carbon.2009.04.026

[40]   T. M. Maria, A. Thomas, S.H. Yu, J.O. Müller, M. Antonietti, A direct synthesis
        of mesoporous carbons with bicontinuous pore morphology from crude plant
        material by hydrothermal carbonization, Chem. Mater. 19 (2007) 4205-4212.
        https://doi.org/10.1021/cm0707408

[41]   Y.S. Hong, X.J. Cui, L.L. Li, K. Li, B. Yu, M. Antonietti, H. Cölfen, From starch
        to metal/carbon hybrid nanostructures: Hydrothermal metal-catalyzed
        carbonization, Adv. Mater. 16 (2004) 1636-1640.
        https://doi.org/10.1002/adma.200400522

[42]   M.A. Lillo-Ródenas, D.C. Amorós, A.L. Solano, Understanding chemical
        reactions between carbons and NaOH and KOH: an insight into the chemical
        activation mechanism, Carbon 41 (2003) 267-275. https://doi.org/10.1016/S0008-
        6223(02)00279-8

[43]   S.G. Badie, A.A. Attia, N.A. Fathy, Modification in adsorption characteristics of
        activated carbon produced by H_3PO_4 under flowing gases, Colloids Surf. A
        Physicochem. Eng. Asp. 299 (2007) 79-87.
        https://doi.org/10.1016/j.colsurfa.2006.11.024

Biomass Based Energy Storage Materials
Materials Research Foundations **78** (2020) 124-142

Materials Research Forum LLC
https://doi.org/10.21741/9781644900871-6

Chapter 6

Cellulose-Derived Electrodes for Energy Storage

Shiqi Li[1], Wenyue Li[2], Zhaoyang Fan[2*]

[1] College of Electronic Information, Hangzhou Dianzi University, Hangzhou, 310018, China

[2] Department of Electrical and Computer Engineering and Nano Tech Center, Texas Tech University, Lubbock, TX 79409, USA

* zhaoyang.fan@ttu.edu

Abstract

This chapter discusses cellulose-derived electrodes and their applications in supercapacitors and batteries. Earth-abundant cellulose can be extracted as microfibers, nanofibers, and nanocrystals with different properties. When functionalized with electroactive materials, cellulose fibers are attractive paper substrates for developing flexible and disposable energy storage devices used in wearable applications. Furthermore, these are renewable precursors to produce porous carbon and carbon nanofibers with a variety of morphologies and properties, which have found broad applications in conventional supercapacitors, high-frequency supercapacitors, Li-ion batteries, Li-S batteries, and other emerging battery technologies.

Keywords

Cellulose, Porous Carbon, Carbon Nanofiber, Flexible Energy Storage, Lithium-Ion Batteries, Lithium-Sulfur Batteries, Supercapacitors, High-Frequency Supercapacitors

Contents

1.   **Introduction**..**125**

2.   **Cellulose based flexible composite electrodes**....................................**126**

3.   **Cellulose carbonization and activation**...**127**

4.   **Cellulose-derived carbon for supercapacitors**...................................**129**

5.   **Cellulose-derived carbon for high-frequency supercapacitors**.........**130**

Biomass Based Energy Storage Materials   Materials Research Forum LLC
Materials Research Foundations **78** (2020) 124-142   https://doi.org/10.21741/9781644900871-6

6.   **Cellulose-derived carbon for lithium-ion batteries** 132

7.   **Cellulose-derived carbon for lithium-sulfur batteries** 132

8.   **Cellulose-derived carbon for other batteries** 135

Conclusion...135

References 136

1.   Introduction

Cellulose, with a chemical formula $(C_6H_{10}O_5)_n$, is the main structural component of the cell walls of plants and certain bacteria [1]. As the most abundant natural polymer and a green resource on earth, it is commonly extracted from woods, cotton, different parts of plants, and agricultural byproducts through mechanical pulping or chemical pulping. These extracted microfibers, being the main component of papers, have a typical lateral dimension at the scale of 10 μm and a length up to several hundreds of micrometers. These can be further purified by removing amorphous components to produce cellulose nanocrystals, rod-like nanoparticles with a length of ~ 100 nm. Cellulose nanofibers, produced from microfibers, have a diameter of tens of nanometers, is another category of interesting nanomaterials. Crosslinked nanofiber network, secreted by certain bacteria in a fermentation process, has also attracted much attention [2].

With their flexibility, hydrophilicity, porosity, lightweight, and low-cost, cellulose microfiber-based papers are becoming attractive substrates for paper-based flexible electronics, supercapacitors, and batteries aiming for wearable applications. For example, the cellulose nanofiber aerogel, with its large specific surface area (SSA) and high porosity, is very appealing to be used as a scaffold to load electrochemically active materials for energy storage. Rod-like cellulose nanocrystals, after cross-linked by introducing functional groups, can also form aerogel for active material loading. Because of its insulating nature, conductive agents need to be introduced into the porous cellulose paper for adoption in flexible electrodes. On the other hand, bacterial cellulose (BC) membrane and cellulose nanocrystal-based membrane might also be used as the insulating separator in batteries or supercapacitors, with the merits of mechanical strength and thermal tolerance. In addition, cellulose is also a good candidate for polymer electrolytes as well as electrode binders. Applications of cellulose and its derivatives for separators and electrolytes have been previously reviewed in detail [3].

On the other hand, due to its earth abundance, low production cost, and appreciable carbon content (40~44 wt%), cellulose is also a renewable resource for producing porous carbon and carbon nanofibers (CNFs), which are widely used as electrode materials in energy storage and conversion systems. Carbon fibers can be directly produced from cellulose microfibers or nanofibers after carbonization process. With porous structures, some of the derived carbon materials have shown their SSA to the level more than 1000 $m^2 g^{-1}$ [4] and hence served as conductive scaffolds for anchoring electroactive materials.

In this chapter, cellulose based flexible paper electrodes for supercapacitors and batteries will be discussed in section 2, and then section 3 will cover cellulose-derived carbon (CDC) nanomaterials, focusing on carbonization and activation processes, which is followed by CDC electrodes for supercapacitors in section 4, high-frequency electrochemical supercapacitors (HF-ECs) in section 5, lithium-on batteries (LIB) in section 6, lithium-sulfur batteries (LSBs) in section 7, and other batteries in section 8.

2.   Cellulose based flexible composite electrodes

Cellulose is attractive as support for active materials to construct composite electrodes, aiming for flexible, disposable and low-cost energy storage [5]. When conductive materials such as carbon allotropes or electronically conductive polymers (ECPs) are used as active components, they can directly form the composite electrodes with cellulose. Otherwise, if insulating transition metal oxides or others used as the electroactive materials, a conductive agent, typically carbon allotropes, will also be necessary in the composite electrodes.

Carbon allotropes, such as graphite, graphene and carbon nanotubes (CNTs), can be directly coated on the surface of cellulose papers to form supercapacitor electrodes. In such a configuration, the cellulose paper is used as a carbon-layer support as well as a separator between the active layers. Filtering carbon nanomaterials into the porous cellulose paper is another method to construct the composite electrode, but an extra layer as the separator will be necessary in this case. Carbon nanomaterials can also be mixed with cellulose pulp and then form the composite paper with a better uniformity. In other examples, cellulose/graphene [6] or cellulose/CNTs aerogel composite electrodes can be obtained through solution mixing to form hydrogel followed by supercritical CO_2 drying, where the cellulose nanofibers are used as spacers, electrolyte reservoirs and hierarchical nanostructure supports in supercapacitors. Through layer-by-layer assembly of CNTs on cross-linked nanofiber aerogels, composite electrodes with a porosity close to 99% with a high mechanical strength can be created [7]. All these electrodes were demonstrated to be promising for flexible supercapacitors.

ECPs, including polyaniline (PANI), polypyrrole (PPy) and poly(3,4-ethylenedioxythiophene) (PEDOT), are attractive electrode materials used in

Biomass Based Energy Storage Materials   Materials Research Forum LLC
Materials Research Foundations **78** (2020) 124-142   https://doi.org/10.21741/9781644900871-6

supercapacitors owing to their inherent fast redox switching and high conductivity. For instance, using a simple soak-and-polymerization method, PPy was uniformly coated onto the surface and inside the bulk of a cellulose paper, which was used for flexible solid-state supercapacitor demonstration [8]. After coating cellulose papers with Au or graphite, PANI was electrodeposited to form flexible electrodes. Polymerizing nanostructured PEDOT onto a cellulose surface or PEDOT nanoparticles into the cellulose paper have also reported. Similarly, cellulose nanofiber aerogel structures, with their large attachable surfaces, can be applied as a scaffold for ECP coatings to achieve a large mass loading. In one study on BC/PPy based electrodes, a large capacitance of 2.43 F cm^{-2} was reported at a discharge current of 2 mA cm^{-2} for the electrode mass loading 11.23 mg cm^{-2} [9]. In another report on PPy/nanofiber based electrodes, a volumetric capacitance of 236 F cm^{-3} was demonstrated for electrodes with an active mass loading as high as 20 mg cm^{-2} [10]. In a most recent development, cellulose/CNTs or graphene/ECP have been integrated together to further enhance both flexibility and conductivity of the electrodes.

Some transition metal oxides can contribute large capacitance thanks to their pseudocapacitive charge storage capability. Through electroless plating, cellulose fiber network was uniformly coated with a Ni layer, on which metal oxides was electrodeposited and then used as flexible supercapacitor electrodes. In another approach, conductive inks, such as CNT or graphene-based inks, were printed onto the cellulose papers to obtain conductive flexible substrates. The porosity of the paper allowed the rapid absorption of these inks. On such flexible conductive substrates, electrochemical materials could be further deposited forming composite electrodes to fabricate flexible batteries [11, 12].

3. Cellulose carbonization and activation

Carbonization of cellulose to produce porous carbon materials with a large SSA has been commonly achieved through two well-developed activation methods, i.e. physical and chemical activation processes. These two methods have their own pros and cons. Physical activation, conducting in an inert environment without introducing any activating agents, can avoid the incorporation of additives/impurities. The obtained carbon products usually have microporous and ultra-microporous structures. However, much higher temperature is necessary to produce these micropores to achieve a large SSA. In contrast, chemical activation with activation agents, is normally conducted at a lower temperature and with a shorter period. Such activated porous carbons generally have larger pore size and higher pore volumes, thus making them comparatively more suitable for liquid electrolytes.

In the chemical activation process, the commonly used activating agents include H_3PO_4, K_2CO_3, KOH and $ZnCl_2$. It was found that the pore size, pore volume, and surface area could be effectively tailored by selecting different activating agents [13]. H_3PO_4-activated carbon possessed the highest SSA and pore volume than those activated by other agents. The carbon monolith attained by pyrolysis of cotton without and with KOH activation possessed specific surface areas of 23.5 and 960 m^2 g^{-1}, respectively [14, 15]. The CNF aerogel derived from BC without activation displayed a SSA of 498 m^2 g^{-1}, while its SSA reached to 1236 m^2 g^{-1} when a suitable amount of KOH was used [16].

Gaseous agents-based activation was also investigated. CNFs obtained from BC by pyrolysis in CO_2 gave much larger SSA than that of argon [17]. Carbons derived from pyrolysis of cellulose paper and BC in NH_3 exhibited remarkable SSA with abundant micropores [18, 19]. Besides creating porous structures and thus increasing their SSA, some agents (such as H_3PO_4 and NH_3) can also introduce phosphor and nitrogen dopants into the obtained carbon matrices [13, 18], thus changing their conductivity and surface polarity, or introducing pseudocapacitive charge storage capability. Steam is another effective activation agent. CDC with an appreciable SSA of ~1300 m^2 g^{-1} and a total pore volume of ~0.6 cm^3 g^{-1} was reported [20] through carbonization at 400°C followed by steam activation up to 900°C with 100°C per second ramping rate. However, such formed activated carbon has a powder or granular shape, while most studies prefer to use CDC fiber network or CNF aerogel that forms a conductive scaffold. Such a self-supported conductive scaffold, after loading active materials, could further be used as a binder-free freestanding electrode.

Considering the production cost and particularly the resulted unique carbon nanostructures, it might also be wiser to directly use raw fibers as the precursor to synthesize hierarchical porous carbon (HPC), instead of starting from pure cellulose. This is because in the raw fibers, hemicellulose, lignin and other components with less crystallinity could not only be transformed into carbon, more importantly, they can also serve as self-templated pore-forming agents to facilitate the formation of HPC structures. In this aspect, other biomasses with dominant cellulose components, in addition to those raw fibers, are also good candidates to produce HPCs. For instance, the main components in rice husk are cellulose (38%), hemicellulose (18%), lignin (22%) and SiO_2. The first three components can be converted into carbon, while SiO_2 nanoparticles serve as a self-templated pore-forming agent that can be etched away after the carbonization process, thus contributing to HPC structure formation. The same is true for bamboo and some other similar biomaterials. For these raw biomasses, hydrothermal or other pre-activation methods are commonly performed for carbon stabilization before high-temperature treatment.

Although microporous and mesoporous carbons with a large SSA are generally required for electrode applications, for some special cases, e.g., high-frequency supercapacitors, micropores must be avoided in order to achieve fast frequency response above hundreds of hertz [21]. In these applications, a rapid plasma pyrolysis process has been proved to be useful in producing cellulose-derived carbon fibers without micropore formation [22].

4. Cellulose-derived carbon for supercapacitors

CDC, with a hierarchical porous structure and a large SSA, can be directly used as electrodes for electrical double layer-based supercapacitors. Carbon meshes derived from pyrolysis of cotton membranes in ammonia gas with a SSA of 1070 m^2 g^{-1} were reported as binder-free electrodes of supercapacitors, which gave a capacitance of 172 F g^{-1} (Fig. 1a-c) [14]. In another study (Fig. 1d-g) [23], raw cotton was first pretreated in NaOH/urea solution for swelling before carbonization. During this pretreatment, the cellulose crystal structure remained unchanged, but the degree of its crystallinity and dimensions of crystallites were modified. This resulted in an increased internal surface area and contributed to development of a porous network with micropores and mesopores formed in the walls of macropores in the subsequent chemical activation process. The synthesized carbon fibers exhibited high specific capacitance of 221.72 and 189.76 F g^{-1} at a current density of 0.3 and 1 A g^{-1}, respectively. Porous nanoplatelets wrapped carbon aerogel structure was also produced in a similar method from bamboo cellulose fibers, giving a high capacitance of 400 F g^{-1} at 5 mV s^{-1} [24]. In comparison, carbon nanofiber aerogels derived from wood-based nanofibrillated cellulose aerogels delivered a capacitance of 140 F g^{-1} at 0.5 A g^{-1} [25].

As a conductive network, CDC fiber meshes are positioned to be a good scaffold structure for anchoring transition metal oxides and other active materials in developing electrodes for pseudocapacitors. For instance, using BC as both a template and also a precursor, nitrogen-doped carbon networks were synthesized through carbonization of polyaniline (PANI) coated BC, on which MnO_2 nanosheets were anchored [26]. In another study, CDC microfiber network was used as scaffold for vertical MoS_2 nanosheets growth, showing its promising potential to be used in high-rate supercapacitors [27].

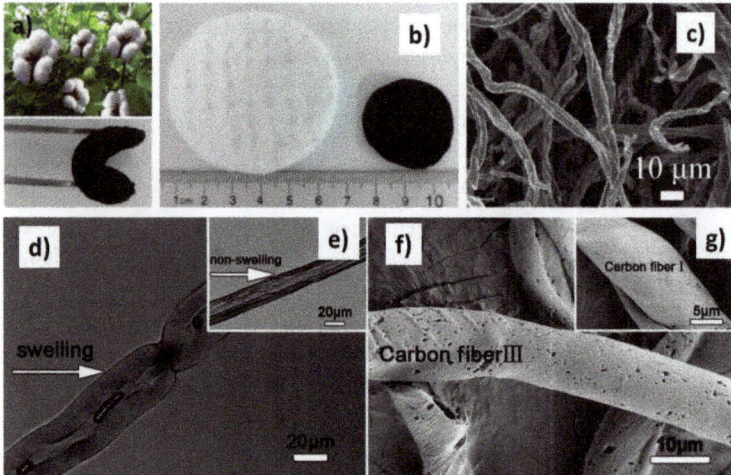

Figure 1. Photos of (a) cotton, (b) the used cotton membrane (white) and the derived carbon fiber mesh (black) after pyrolysis, (c) Scanning electron microscope (SEM) image of carbon fibers. Optical microscope images of (d) the fiber in NaOH/urea aqueous solutions, and (e) native fiber. SEM image of carbon fiber of (f) with swelling and calcination and (g) only calcination. Reproduced with permission from Ref.[14] for (a-c), Copyright 2019, Elsevier, and from Ref. [23] for (d-g), Copyright 2016, American Chemical Society.

5. Cellulose-derived carbon for high-frequency supercapacitors

Conventional supercapacitors, with electrodes having abundant mesopores and micropores, have a frequency response below 1 Hz. There are strong efforts in developing high-frequency electrochemical supercapacitors (HF-ECs) working at hundreds or kilo-hertz frequencies for line-frequency alternating current filtering or pulse power generation and storage [21]. Through rapid plasma pyrolysis, CDC fiber network and aerogel, with the absence of micropores, have been developed and found suitable to be employed as electrodes of HF-ECs [28].

Fig. 2 shows the results of using BC-derived cross-linked CNF aerogel as electrodes for HF-ECs [22, 29]. After rapid plasma pyrolysis of BC aerogel, cross-linked CNF was obtained (Fig. 2b) with a small SSA of 57.5 m^2 g^{-1} and a minimum pore size of 3.8 nm. The complex-plane impedance spectra for electrodes produced respectively by plasma rapid pyrolysis and thermal pyrolysis methods are compared in Fig. 2c, which indicate very slow frequency response from electrodes derived by thermal pyrolysis. As in Fig. 2d, the aqueous cell with 20 μm thick electrode shows that the absolute value of the phase angle stays above 80° with frequency increasing to a few hundred Hz. In particular, the 120 Hz phase angle is around −82° and the characteristic frequency when the phase angle reaches −45° is close to 3.3 kHz. The areal capacitances of a single electrode at 120 Hz were 1.51, 2.98, and 4.50 mF cm^{-2} for 10, 20, and 60 μm electrode, respectively. In addition, for the 10 μm electrode, a phase angle of −80°, and an areal capacitance of 0.51 mF cm^{-2} at 120 Hz were also achieved with an operation window of more than 3V in a cell with an organic electrolyte.

In other studies, cellulose tissue paper or filter paper was treated in a plasma-enhanced chemical vapor deposition system for a short time, with CH_4 and H_2 as precursor gases. Cellulose paper carbonization and vertical graphene growth [30, 31] were carried out simultaneously. Such formed graphene/CDC freestanding electrodes were found to be suitable for HF-ECs with large capacitance and kilo-hertz frequency response [32]. They have been used to demonstrate the applications of HF-ECs for voltage ripple filtering in AC/DC converter and pulse power storage and smoothing in energy harvesters [33, 34].

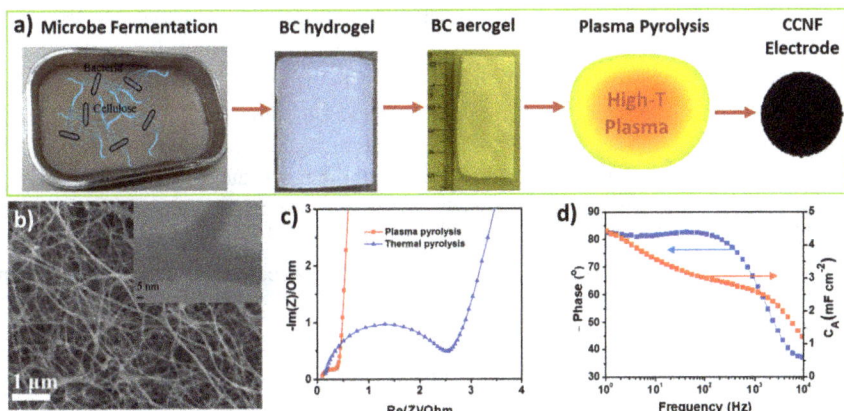

Figure 2. (a) Schematic showing the BC-derived cross-linked CNF fabrication process. (b) The SEM image of cross-linked CNF with macro-pores. The cross-link between CNFs can be observed from the inset transmission electron microscope (TEM) image. (c) Difference in their complex impedance spectra for CNF electrodes fabricated by plasma pyrolysis and by thermal pyrolysis respectively. (h) Dependence of the phase angle and the electrode capacitance on frequency for a CNF electrode with a thickness of 20 μm. Reproduced with permission from Ref.[22] Copyright 2017, Elsevier.

Biomass Based Energy Storage Materials   Materials Research Forum LLC
Materials Research Foundations **78** (2020) 124-142   https://doi.org/10.21741/9781644900871-6

6.   Cellulose-derived carbon for lithium-ion batteries

In conventional LIBs, graphite is commonly used as the anode material, which has a low theoretical capacity of only 372 mAh g^{-1}. Different CDC materials have been investigated aiming for a larger anode capacity. Carbon nanofibers with a specific surface area of 1236 m^2 g^{-1} were derived from BC [16]. When applied as the anode material for a LIB, it delivered a specific capacity of 858 mAh g^{-1} after 100 cycles at 100 mA g^{-1} and maintained a capacity of 325 mAh g^{-1} at a current density of 400 mA g^{-1}. Its excellent performance was ascribed to the cross-linked structure and abundance of micropores and mesopores in the activated BC-derived CNFs.

Other high-capacity anode materials can also be loaded on CDC to further increase the anode capacity. For instance, Fe_2O_3 nanoparticles, with a high theoretical capacity of 925 mAh g^{-1}, were anchored on BC-derived carbon nanofiber aerogel [35], which delivered a high reversible capacity of 754 mAh g^{-1} after 100 cycles at 100 mA g^{-1}. In another study, amorphous tin oxide gel film was first deposited on a natural cellulose paper and then the composite was transformed to carbon nanofibers coated with crystalline SnO_2 nanoparticles after pyrolysis [36]. The hierarchical structure, composed of 3-dimentional carbon microfiber network and ultrafine SnO_2 nanoparticles, facilitated the electrode-electrolyte contact, promoted the electron transfer as well as Li^+ transport and meanwhile alleviated the severe volume change during charging/discharging. As an anode of LIBs, it delivered a capacity of 623 mAh g^{-1} after 120 cycles at 0.2 C. Through a similar process, Si/cellulose composite was also prepared and then transformed to Si@ SiO_2@C composite [37]. In this architecture, carbon derived from cellulose served as not only a highly conductive network for electron transportation but also as buffer for the huge volume change of Si during charging/discharging process. The rationally designed electrode was demonstrated to have excellent performances, delivering a capacity of 1071 mAh g^{-1} at the current density of 420 mA g^{-1} after 200 cycles, ascribing to the 3-D interconnected architecture of the carbon nanofibers inherited from the cellulose nanofibers [38].

7.   Cellulose-derived carbon for lithium-sulfur batteries

LSBs, with sulfur as the cathode and lithium as the anode, have been considered as a promising next-generation battery technology owing to their high theoretical energy densities (2600 Wh Kg^{-1}). However, several problems have hindered their practical applications. These include the poor electronic and ionic conductivities of sulfur, the notorious shuttle effect stemming from the dissolution and diffusion of lithium polysulfides (Li_2S_x, $4 \leq x \leq 8$), and the large volume variation resulting from density difference between sulfur and lithium sulfide.

Biomass Based Energy Storage Materials | Materials Research Forum LLC
Materials Research Foundations **78** (2020) 124-142 | https://doi.org/10.21741/9781644900871-6

Carbon materials are usually adopted to host sulfur aiming to enhance electronic conductivity and alleviate the volume fluctuation in the sulfur cathode, and meanwhile confine the sulfur species, thereby improving the electrochemical performance. CDC usually retains the morphology of cellulose with carbon fibers interconnected with each other, which could facilitate electron and Li^+ transportation in the formed sulfur/CDC composite electrode. For instance, the conventional filter paper was pyrolyzed into a carbon fiber film which was then used as an interlayer for LSBs (Fig. 3) [39]. Such an interlayer was found to dramatically improve the cell capacity since it can trap and thus prevent the diffusion of those soluble polysulfides. The interlay, acting as a second current collector, also dramatically enhances the charge-discharge rate capability of LSBs. The resulted LSBs delivered a capacity more than 830 mAh g^{-1} and an efficiency of more than 97% at 0.2 C rate.

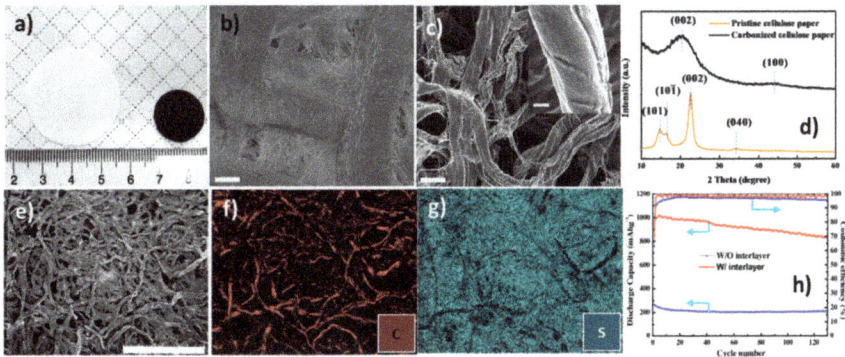

Figure 3. (a) Photos of the cellulose filter paper (white) and after the carbonization (black). (b) SEM image of the filter paper (scale bar: 10 μm). (c) SEM image of the carbonized paper (scale bar: 10 μm). A higher magnification SEM image of an individual carbon fiber is shown in the inset (scale bar: 1μm). (d) X-ray diffraction patterns of the cellulose paper and the carbonized paper. SEM and Energy-dispersive X-ray spectroscopy (EDS) characterization of the interlayer after 130 charge-discharge cycles at 0.2 C: (e) SEM image of the interlayer after cycling (scale bar: 200μm), and the corresponding C mapping (f) and S mapping (g). (h) Cycling performance of Li-S cells without and with interlayer at 0.2 C. Reproduced with permission from Ref. [39], Copyright 2017, Elsevier.

Biomass Based Energy Storage Materials   Materials Research Forum LLC
Materials Research Foundations **78** (2020) 124-142   https://doi.org/10.21741/9781644900871-6

It is known that micropores are superior in confining polysulfides while inferior in facilitating Li^+ transportation compared with mesopores and macropores [40]. Therefore, micropores would contribute to high initial capacity and mesopores render stable cyclability [41]. Hence, it is critical to tailor the surface area and pore size distribution in CDC. In one study, bamboo leaves, consisting of SiO_2 nanoparticles, cellulose, semi-cellulose, and lignin, were converted into hierarchical porous carbon *via* a carbonization process followed by an etching step to remove the residues of SiO_2 nanoparticles [42]. With the attained carbon as the sulfur host in the cathode, LSB delivered a high initial capacity of 1487 mAh g^{-1} at 0.05 C and retained a capacity of 707 mAh g^{-1} after 200 cycles at 1 C. Similarly, porous carbon derived from rice husk [41], flexible carbon monolith with a large SSA converted from cotton textile [15], and carbon slices produced from bark of plane trees [43], were also reported and employed as host for sulfur cathode material. Specifically, the aforementioned carbon slices/sulfur cathode showed an initial specific capacity of 1159 mAh g^{-1} at 0.2 A g^{-1} with a high coulombic efficiency of ~98%.

The weak interaction between nonpolar carbon and polar polysulfides leads to the physical confinement of polysulfides not very effective. Heteroatom-doped carbon with modified polarity was discovered to be more effective in confining polysulfides in the cathode through chemical binding [44, 45]. Porous carbon doped with N and S was fabricated by pyrolysis of cellulose nanocrystals decorated with polyrhodanine, and the resulted LSBs demonstrated excellent cycling stability and rate capability [46].

In one study, hierarchical porous carbon derived from lotus seedpod shells showed high cellulose content [47]. To enhance interactions between polysulfides and the carbon host, ionic surfactants were grafted on the surface of the derived hierarchical porous carbon. With both physical trapping and chemical binding effects, dissolution and diffusion of polysulfides in LSBs were inhibited. Calculation results suggested that O or N containing functional groups in carbon materials could strongly bind polysulfides [44, 48]. We attained N and O dual-doped carbon nanoribbon aerogels with a surface area of 773 m^2 g^{-1} by pyrolysis of BC in ammonia [48, 49]. LSBs based on the carbon nanoribbon aerogel delivered a capacity as high as 943 mA h g^{-1} even when a sulfur loading as high as 6.4 mg cm^{-2} and a sulfur content as large as 90% were used, corresponding to an areal capacity of 5.9 mAh cm^{-2}. The outstanding cell performance was ascribed to the dual confinements of active materials by physical localization and chemical bonding, together with the 3D interconnected web structure of the carbon nanoribbon aerogel.

Biomass Based Energy Storage Materials   Materials Research Forum LLC
Materials Research Foundations **78** (2020) 124-142   https://doi.org/10.21741/9781644900871-6

8. Cellulose-derived carbon for other batteries

CDC has also been investigated as an electrode material for other batteries such as Zn-air battery, Na-ion battery and so on. The N-doped CNF aerogel derived from BC pyrolysis in NH_3, when applied as a cathode catalyst in Zn-air battery, exhibited a high voltage of 1.34 V at a discharge current density of 1.0 mA cm^{-2}, while it remained to be 1.25 V even when the discharge current density increased to 10 mA cm^{-2}. These results are comparable to those obtained using the state-of-the-art Pt/C catalyst [19]. The excellent performance was attributed to easy ion diffusion in the gel and rapid electron transport in N-doped CNF. Moreover, the doped nitrogen also is expected also to serve as active electrocatalysis sites for oxygen reduction in order to facilitate the charge transfer. BC was also used as a skeleton to anchor another polymer for producing heavily doped CNF. Pyrrole was coated on BC and then the hydrogel was transformed to N-doped CNF by hydrothermal pretreatment and the subsequent pyrolysis [50], which showed much better electrochemical performance than that derived from pure BC.

When applied as anodes of Na-ion batteries, hard CNFs derived from cellulose nanofibers extracted from bleached pulp demonstrated a reversible capacity of 255 mAh g^{-1} at a current density of 40 mA g^{-1}. These also exhibited a good rate capability with a capacity of 85 mAh g^{-1} at a discharge current of 2 A g^{-1}. A capacity of 176 mAh g^{-1} was maintained after 600 cycles at 200 mA g^{-1} [51]. In another study [52], cellulose-derived hard carbon, produced by first treatment at 275 °C in air and then 1300–1500 °C in argon, was found to deliver superior reversible sodiation capacity (353 mA h g^{-1}). The low temperature pretreatment was found to be critical in controlling dehydration and cross-linkage degrees of cellulose, which gave rise to the nanoscale structure of hard carbons. These hard carbons delivered capacity values of 523 and 290 mAh g^{-1} in Li and K batteries, respectively.

Cellulose nanocrystals could be transformed into crystalline ordered carbon at a relatively low carbonization temperature [53]. When applied as the anode in Na-ion batteries, the ordered carbon delivered an initial capacity of 340 mAh g^{-1} at a current density of 100 mA g^{-1} and maintained 88.2% capacity after 400 cycles. This outstanding performance was attributed to the large interlayer space in the crystalline ordered carbon for the Na^+ intercalation and the proper pore size for electrolyte adsorbing. The crystalline ordered structure could also provide fast electronic and ion transportation pathways.

Conclusion

Cellulose, as an earth-abundant and renewable source, has already found many potential applications in energy storage and conversion systems. As paper substrates, they can be

Biomass Based Energy Storage Materials
Materials Research Foundations **78** (2020) 124-142

Materials Research Forum LLC
https://doi.org/10.21741/9781644900871-6

functionalized with other active materials to construct flexible and disposable energy storage used in wearable electronics. With its appreciable carbon content, cellulose has also been extensively used to produce porous carbon and carbon nanofibers, which have been investigated as electrodes for a variety of electrochemical energy storages from conventional supercapacitors, high-frequency supercapacitors, to lithium ion batteries, lithium sulfur batteries, and other emerging battery technologies. All these indicate that cellulose could play critical roles in electrochemical energy storage systems.

References

[1]   H.A. Khalil, A. Bhat, A.I. Yusra, Green composites from sustainable cellulose nanofibrils: A review, Carbohydr. Polym. 87 (2012) 963-979. https://doi.org/10.1016/j.carbpol.2011.08.078

[2]   I. Siró, D. Plackett, Microfibrillated cellulose and new nanocomposite materials: A review, Cellulose 17 (2010) 459-494. https://doi.org/10.1007/s10570-010-9405-y

[3]   L. Jabbour, R. Bongiovanni, D. Chaussy, C. Gerbaldi, D. Beneventi, Cellulose-based Li-ion batteries: A review, Cellulose 20 (2013) 1523-1545. https://doi.org/10.1007/s10570-013-9973-8

[4]   V. Gupta, P. Carrott, R. Singh, M. Chaudhary, S. Kushwaha, Cellulose: A review as natural, modified and activated carbon adsorbent, Bioresource Technol. 216 (2016) 1066-1076. https://doi.org/10.1016/j.biortech.2016.05.106

[5]   Z. Wang, P. Tammela, M. Strømme, L. Nyholm, Cellulose-based supercapacitors: material and performance considerations, Adv. Energy Mater. 7 (2017) 1700130. https://doi.org/10.1002/aenm.201700130

[6]   K. Gao, Z. Shao, J. Li, X. Wang, X. Peng, W. Wang, F. Wang, Cellulose nanofiber–graphene all solid-state flexible supercapacitors, J. Mater. Chem. A 1 (2013) 63-67. https://doi.org/10.1039/C2TA00386D

[7]   M. Hamedi, E. Karabulut, A. Marais, A. Herland, G. Nyström, L. Wågberg, Nanocellulose aerogels functionalized by rapid layer-by-layer assembly for high charge storage and beyond, Angew. Chem. Int. Ed. 125 (2013) 12260-12264. https://doi.org/10.1002/ange.201305137

[8]   L. Yuan, B. Yao, B. Hu, K. Huo, W. Chen, J. Zhou, Polypyrrole-coated paper for flexible solid-state energy storage, Energy Environ. Sci. 6 (2013) 470-476. https://doi.org/10.1039/c2ee23977a

[9] S. Li, D. Huang, J. Yang, B. Zhang, X. Zhang, G. Yang, M. Wang, Y. Shen, Freestanding bacterial cellulose–polypyrrole nanofibres paper electrodes for advanced energy storage devices, Nano Energy 9 (2014) 309-317. https://doi.org/10.1016/j.nanoen.2014.08.004

[10] Z. Wang, P. Tammela, P. Zhang, M. Strømme, L. Nyholm, High areal and volumetric capacity sustainable all-polymer paper-based supercapacitors, J. Mater. Chem. A 2 (2014) 16761-16769. https://doi.org/10.1039/C4TA03724C

[11] L. Hu, J.W. Choi, Y. Yang, S. Jeong, F. La Mantia, L.F. Cui, Y. Cui, Highly conductive paper for energy-storage devices, Proc. Natl. Acad. Sci. 106 (2009) 21490-21494. https://doi.org/10.1073/pnas.0908858106

[12] S. Jeong, N. Böckenfeld, A. Balducci, M. Winter, S. Passerini, Natural cellulose as binder for lithium battery electrodes, J. Power Sources 199 (2012) 331-335. https://doi.org/10.1016/j.jpowsour.2011.09.102

[13] H. Wu, Y. Deng, J. Mou, Q. Zheng, F. Xie, E. Long, C. Xu, D. Lin, Activator-induced tuning of micromorphology and electrochemical properties in biomass carbonaceous materials derived from mushroom for lithium-sulfur batteries, Electrochim Acta (2017). https://doi.org/10.1016/j.electacta.2017.05.026

[14] S. Li, Z. Fan, Nitrogen-doped carbon mesh from pyrolysis of cotton in ammonia as binder-free electrodes of supercapacitors, Microporous Mesoporous Mater. 274 (2019) 313-317. https://doi.org/10.1016/j.micromeso.2018.09.002

[15] Z. Gao, Y. Zhang, N. Song, X. Li, Towards flexible lithium-sulfur battery from natural cotton textile, Electrochim Acta 246 (2017). https://doi.org/10.1016/j.electacta.2017.06.069

[16] W. Wang, Y. Sun, B. Liu, S. Wang, M. Cao, Porous carbon nanofiber webs derived from bacterial cellulose as an anode for high performance lithium ion batteries, Carbon 91 (2015) 56-65. https://doi.org/10.1016/j.carbon.2015.04.041

[17] L.F. Chen, Z.H. Huang, H.W. Liang, W.T. Yao, Z.Y. Yu, S.H. Yu, Flexible all-solid-state high-power supercapacitor fabricated with nitrogen-doped carbon nanofiber electrode material derived from bacterial cellulose, Energy Environ. Sci. 6 (2013) 3331-3338. https://doi.org/10.1039/c3ee42366b

[18] W. Luo, B. Wang, C.G. Heron, M.J. Allen, J. Morre, C.S. Maier, W.F. Stickle, X. Ji, Pyrolysis of cellulose under ammonia leads to nitrogen-doped nanoporous

carbon generated through methane formation, Nano Lett. 14 (2014) 2225-2229.
https://doi.org/10.1021/nl500859p

[19]   H.W. Liang, Z.Y. Wu, L.F. Chen, C. Li, S.H. Yu, Bacterial cellulose derived
       nitrogen-doped carbon nanofiber aerogel: An efficient metal-free oxygen reduction
       electrocatalyst for zinc-air battery, Nano Energy 11 (2015) 366-376.
       https://doi.org/10.1016/j.nanoen.2014.11.008

[20]   E. Lorenc-Grabowska, P. Rutkowski, High basicity adsorbents from solid residue
       of cellulose and synthetic polymer co-pyrolysis for phenol removal: kinetics and
       mechanism, Appl. Surf. Sci. 316 (2014) 435-442.
       https://doi.org/10.1016/j.apsusc.2014.08.024

[21]   Z. Fan, N. Islam, S.B. Bayne, Towards kilohertz electrochemical capacitors for
       filtering and pulse energy harvesting, Nano Energy 39 (2017) 306-320.
       https://doi.org/10.1016/j.nanoen.2017.06.048

[22]   N. Islam, S. Li, G. Ren, Y. Zu, J. Warzywoda, S. Wang, Z. Fan, High-frequency
       electrochemical capacitors based on plasma pyrolyzed bacterial cellulose aerogel
       for current ripple filtering and pulse energy storage, Nano Energy 40 (2017) 107-
       114. https://doi.org/10.1016/j.nanoen.2017.08.015

[23]   Y. Liu, Z. Shi, Y. Gao, W. An, Z. Cao, J. Liu, Biomass-swelling assisted synthesis
       of hierarchical porous carbon fibers for supercapacitor electrodes, ACS Appl.
       Mater. Interfaces 8 (2016) 28283-28290. https://doi.org/10.1021/acsami.5b11558

[24]   X. Yang, B. Fei, J. Ma, X. Liu, S. Yang, G. Tian, Z. Jiang, Porous nanoplatelets
       wrapped carbon aerogels by pyrolysis of regenerated bamboo cellulose aerogels as
       supercapacitor electrodes, Carbohydr. Polym. 180 (2018) 385-392.
       https://doi.org/10.1016/j.carbpol.2017.10.013

[25]   S.C. Li, B.C. Hu, Y.W. Ding, H.W. Liang, C. Li, Z.Y. Yu, Z.Y. Wu, W.S. Chen,
       S.H. Yu, Wood-derived ultrathin carbon nanofiber aerogels, Angew. Chem. Int.
       Ed. 130 (2018) 7203-7208. https://doi.org/10.1002/ange.201802753

[26]   C. Long, D. Qi, T. Wei, J. Yan, L. Jiang, Z. Fan, Nitrogen-doped carbon networks
       for high energy density supercapacitors derived from polyaniline coated bacterial
       cellulose, Adv. Funct. Mater. 24 (2014) 3953-3961.
       https://doi.org/10.1002/adfm.201304269

[27] N. Islam, S. Wang, J. Warzywoda, Z. Fan, Fast supercapacitors based on vertically oriented MoS2 nanosheets on plasma pyrolyzed cellulose filter paper, J. Power Sources 400 (2018) 277-283. https://doi.org/10.1016/j.jpowsour.2018.08.049

[28] W. Li, N. Islam, G. Ren, S. Li, Z. Fan, AC-filtering supercapacitors based on edge oriented vertical graphene and cross-linked carbon nanofiber, Materials 12 (2019) 604. https://doi.org/10.3390/ma12040604

[29] N. Islam, M.N.F. Hoque, Y. Zu, S. Wang, Z. Fan, Carbon nanofiber aerogel converted from bacterial cellulose for kilohertz AC-supercapacitors, MRS Adv. 3 (2018) 855-860. https://doi.org/10.1557/adv.2018.139

[30] G. Ren, X. Pan, S. Bayne, Z. Fan, Kilohertz ultrafast electrochemical supercapacitors based on perpendicularly-oriented graphene grown inside of nickel foam, Carbon 71 (2014) 94-101. https://doi.org/10.1016/j.carbon.2014.01.017

[31] G. Ren, M.N.F. Hoque, X. Pan, J. Warzywoda, Z. Fan, Vertically aligned VO_2 (B) nanobelt forest and its three-dimensional structure on oriented graphene for energy storage, J. Mater. Chem. A 3 (2015) 10787-10794. https://doi.org/10.1039/C5TA01900A

[32] G. Ren, S. Li, Z.-. Fan, M.N.F. Hoque, Z. Fan, Ultrahigh-rate supercapacitors with large capacitance based on edge oriented graphene coated carbonized cellulous paper as flexible freestanding electrodes, J. Power Sources 325 (2016) 152-160. https://doi.org/10.1016/j.jpowsour.2016.06.021

[33] N. Islam, M.N.F. Hoque, W. Li, S. Wang, J. Warzywoda, Z. Fan, Vertically edge-oriented graphene on plasma pyrolyzed cellulose fibers and demonstration of kilohertz high-frequency filtering electrical double layer capacitors, Carbon 141 (2019) 523-530. https://doi.org/10.1016/j.carbon.2018.10.012

[34] N. Islam, J. Warzywoda, Z. Fan, Edge-oriented graphene on carbon nanofiber for high-frequency supercapacitors, Nano-Micro Lett. 10 (2018) 9. https://doi.org/10.1007/s40820-017-0162-4

[35] Y. Wan, Z. Yang, G. Xiong, H. Luo, A general strategy of decorating 3D carbon nanofiber aerogels derived from bacterial cellulose with nano-Fe_3O_4 for high-performance flexible and binder-free lithium-ion battery anodes, J. Mater. Chem. A 3 (2015) 15386-15393. https://doi.org/10.1039/C5TA03688G

[36] W. Mengya, L. Shun, Z. Yiming, H. Jianguo, Hierarchical SnO_2/carbon nanofibrous composite derived from cellulose substance as anode material for lithium-ion batteries, Chemistry 21 (2015) 16195-16202. https://doi.org/10.1002/chem.201502833

[37] D. Shen, C. Huang, L. Gan, J. Liu, Z. Gong, M. Long, Rational design of Si@SiO2/C composite using sustainable cellulose as carbon resource for anode in lithium-ion batteries, ACS Appl. Mater. Interfaces 10 (2018) 7946-7954. https://doi.org/10.1021/acsami.7b16724

[38] J.M. Kim, V. Guccini, K.D. Seong, J. Oh, G. Salazar-Alvarez, Y. Piao, Extensively interconnected silicon nanoparticles via carbon network derived from ultrathin cellulose nanofibers as high performance lithium ion battery anodes, Carbon (2017). https://doi.org/10.1016/j.carbon.2017.03.028

[39] S. Li, G. Ren, M.N.F. Hoque, Z. Dong, J. Warzywoda, Z. Fan, Carbonized cellulose paper as an effective interlayer in lithium-sulfur batteries, Appl. Surf. Sci.396 (2016) 637–643. https://doi.org/10.1016/j.apsusc.2016.10.208

[40] L. Yu, N. Brun, K. Sakaushi, J. Eckert, M.M. Titirici, Hydrothermal nanocasting: synthesis of hierarchically porous carbon monoliths and their application in lithium–sulfur batteries, Carbon 61 (2013) 245-253. https://doi.org/10.1016/j.carbon.2013.05.001

[41] M.K. Rybarczyk, H.-J. Peng, C. Tang, M. Lieder, Q. Zhang, M.M. Titirici, Porous carbon derived from rice husks as sustainable bioresources: insights into the role of micro-/mesoporous hierarchy in hosting active species for lithium–sulphur batteries, Green Chem. 18 (2016) 5169-5179. https://doi.org/10.1039/C6GC00612D

[42] Y. Li, L. Wang, B. Gao, X. Li, Q. Cai, Q. Li, X. Peng, K. Huo, P.K. Chu, Hierarchical porous carbon materials derived from self-template bamboo leaves for lithium–sulfur batteries, Electrochim Acta 229 (2017) 352-360. https://doi.org/10.1016/j.electacta.2017.01.166

[43] J. Xu, K. Zhou, F. Chen, W. Chen, X. Wei, X.-W. Liu, J. Liu, Natural integrated carbon architecture for rechargeable lithium–sulfur batteries, ACS Sustain. Chem. Eng. 4 (2016) 666-670. https://doi.org/10.1021/acssuschemeng.5b01258

[44] S. Li, T. Mou, G. Ren, J. Warzywoda, B. Wang, Z. Fan, Confining sulfur species in cathodes of lithium–sulfur batteries: insight into nonpolar and polar matrix

Biomass Based Energy Storage Materials
Materials Research Forum LLC
Materials Research Foundations **78** (2020) 124-142
https://doi.org/10.21741/9781644900871-6

surfaces, ACS Energy Lett. 1 (2016) 481-489.
https://doi.org/10.1021/acsenergylett.6b00182

[45]   G. Ren, S. Li, Z.-X. Fan, J. Warzywoda, Z. Fan, Soybean-derived hierarchical
       porous carbon with large sulfur loading and sulfur content for high-performance
       lithium–sulfur batteries, J. Mater. Chem. A 4 (2016) 16507-16515.
       https://doi.org/10.1039/C6TA07446D

[46]   P. Quan, T. Juntao, H. He, L. Xiao, H. Connor, K.C. Tam, L.F. Nazar, A nitrogen
       and sulfur dual-doped carbon derived from polyrhodanine@cellulose for advanced
       lithium-sulfur batteries, Adv. Mater. 27 (2015) 6021.
       https://doi.org/10.1002/adma.201502467

[47]   M. Chen, S. Jiang, S. Cai, X. Wang, K. Xiang, Z. Ma, P. Song, A.C. Fisher,
       Hierarchical porous carbon modified with ionic surfactants as efficient sulfur hosts
       for the high-performance lithium-sulfur batteries, Chem. Eng. J. 313 (2017) 404-
       414. https://doi.org/10.1016/j.cej.2016.12.081

[48]   S. Li, T. Mou, G. Ren, J. Warzywoda, Z. Wei, B. Wang, Z. Fan, Gel based sulfur
       cathodes with a high sulfur content and large mass loading for high-performance
       lithium–sulfur batteries, J. Mater. Chem. A 5 (2017) 1650-1657.
       https://doi.org/10.1039/C6TA09841J

[49]   S. Li, J. Warzywoda, S. Wang, G. Ren, Z. Fan, Bacterial cellulose derived carbon
       nanofiber aerogel with lithium polysulfide catholyte for lithium–sulfur batteries,
       Carbon 124 (2017) 212-218. https://doi.org/10.1016/j.carbon.2017.08.062

[50]   Y. Huang, L. Wang, L. Lu, M. Fan, F. Yuan, B. Sun, J. Qian, Q. Hao, D. Sun,
       Preparation of bacterial cellulose based nitrogen-doped carbon nanofibers and
       their applications in the oxygen reduction reaction and sodium–ion battery, New J.
       Chem. 42 (2018) 7407-7415. https://doi.org/10.1039/C8NJ00708J

[51]   W. Luo, J. Schardt, C. Bommier, B. Wang, J. Razink, J. Simonsen, X. Ji, Carbon
       nanofibers derived from cellulose nanofibers as a long-life anode material for
       rechargeable sodium-ion batteries, J. Mater. Chem.A 1 (2013) 10662-10666.
       https://doi.org/10.1039/c3ta12389h

[52]   H. Yamamoto, S. Muratsubaki, K. Kubota, M. Fukunishi, H. Watanabe, J. Kim, S.
       Komaba, Synthesizing higher-capacity hard-carbons from cellulose for Na- and K-
       ion batteries, J. Mater. Chem. A 6 (2018) 16844-16848
       https://doi.org/10.1039/C8TA05203D

Biomass Based Energy Storage Materials   Materials Research Forum LLC
Materials Research Foundations **78** (2020) 124-142   https://doi.org/10.21741/9781644900871-6

[53]   H. Zhu, S. Fei, L. Wei, S. Zhu, M. Zhao, B. Natarajan, J. Dai, L. Zhou, X. Ji, R.S.
       Yassar, Low temperature carbonization of cellulose nanocrystals for high
       performance carbon anode of sodium-ion batteries, Nano Energy 33 (2017) 37-44.
       https://doi.org/10.1016/j.nanoen.2017.01.021

Keyword Index

Activated Carbon 111

Bamboo Stick 111
Biochar.. 1
Bioelectrodes 1
Biofilm.. 1
Biomass 50

Capacitance.. 1
Carbon Nanofiber 124
Cellulose 124
Charge.. 1
Composite Materials........................... 91
Consituents 21

Electrode... 111
Energy Density 50
Energy Storage 50, 91

Flexible Energy Storage 124

Green Energy 1

High-Frequency Supercapacitors 124

Initiation Methods.............................. 21

Lignin... 91
Lithium-Ion Batteries 124
Lithium-Sulfur Batteries................... 124

Natural Precursors 21
Nature Procured Carbons 21

Porous Carbon 50, 124
Power Density 50

Specific Capacitance 111
Structural-Characteristics
Interrelationship 21

Supercapacitor..........21, 50, 91, 111, 124

About the Editors

Dr. Inamuddin is currently working as Assistant Professor in the Chemistry Department, Faculty of Science, King Abdulaziz University, Jeddah, Saudi Arabia. He is a permanent faculty member (Assistant Professor) at the Department of Applied Chemistry, Aligarh Muslim University, Aligarh, India. He obtained Master of Science degree in Organic Chemistry from Chaudhary Charan Singh (CCS) University, Meerut, India, in 2002. He received his Master of Philosophy and Doctor of Philosophy degrees in Applied Chemistry from Aligarh Muslim University (AMU), India, in 2004 and 2007, respectively. He has extensive research experience in multidisciplinary fields of Analytical Chemistry, Materials Chemistry, and Electrochemistry and, more specifically, Renewable Energy and Environment. He has worked on different research projects as project fellow and senior research fellow funded by University Grants Commission (UGC), Government of India, and Council of Scientific and Industrial Research (CSIR), Government of India. He has received Fast Track Young Scientist Award from the Department of Science and Technology, India, to work in the area of bending actuators and artificial muscles. He has completed four major research projects sanctioned by University Grant Commission, Department of Science and Technology, Council of Scientific and Industrial Research, and Council of Science and Technology, India. He has published 171 research articles in international journals of repute and eighteen book chapters in knowledge-based book editions published by renowned international publishers. He has published 105 edited books with Springer (U.K.), Elsevier, Nova Science Publishers, Inc. (U.S.A.), CRC Press Taylor & Francis Asia Pacific, Trans Tech Publications Ltd. (Switzerland), IntechOpen Limited (U.K.), Wiley-Scrivener, (U.S.A.) and Materials Research Forum LLC (U.S.A). He is a member of various journals' editorial boards. He is also serving as Associate Editor for journals (Environmental Chemistry Letter, Applied Water Science and Euro-Mediterranean Journal for Environmental Integration, Springer-Nature), Frontiers Section Editor (Current Analytical Chemistry, Bentham Science Publishers), Editorial Board Member (Scientific Reports-Nature), Editor (Eurasian Journal of Analytical Chemistry), and Review Editor (Frontiers in Chemistry, Frontiers, U.K.) He is also guest-editing various special thematic special issues to the journals of Elsevier, Bentham Science Publishers, and John Wiley & Sons, Inc. He has attended as well as chaired sessions in various international and national conferences. He has worked as a Postdoctoral Fellow, leading a research team at the Creative Research Initiative Center for Bio-Artificial Muscle, Hanyang University, South Korea, in the field of renewable energy, especially biofuel cells. He has also worked as a Postdoctoral Fellow at the Center of Research Excellence in Renewable Energy, King Fahd University of Petroleum and Minerals, Saudi Arabia, in the field of polymer electrolyte membrane fuel cells and computational fluid dynamics of polymer electrolyte membrane fuel cells. He is a life member of the Journal of the Indian Chemical Society. His research interest includes ion exchange materials, a sensor for heavy metal ions, biofuel cells, supercapacitors and bending actuators.

Dr. Rajender Boddula is currently working with Chinese Academy of Sciences-President's International Fellowship Initiative (CAS-PIFI) at National Center for Nanoscience and Technology (NCNST, Beijing). He obtained Master of Science in Organic Chemistry from Kakatiya University, Warangal, India, in 2008. He received his Doctor of Philosophy in Chemistry with the highest honours in 2014 for the work entitled "Synthesis and Characterization of Polyanilines for Supercapacitor and Catalytic Applications" at the CSIR-Indian Institute of Chemical Technology (CSIR-IICT) and Kakatiya University (India). Before joining National Center for Nanoscience and Technology (NCNST) as CAS-PIFI research fellow, China, worked as senior research associate and Postdoc at National Tsing-Hua University (NTHU, Taiwan) respectively in the fields of bio-fuel and CO_2 reduction applications. His academic honors include University Grants Commission National Fellowship and many merit scholarships, study-abroad fellowships from Australian Endeavour Research Fellowship, and CAS-PIFI. He has published many scientific articles in international peer-reviewed journals and has authored around twenty book chapters, and he is also serving as an editorial board member and a referee for reputed international peer-reviewed journals. He has published edited books with Springer (UK), Elsevier, Materials Research Forum LLC (USA), Wiley-Scrivener, (U.S.A.) and CRC Press Taylor & Francis group. His specialized areas of research are energy conversion and storage, which include sustainable nanomaterials, graphene, polymer composites, heterogeneous catalysis for organic transformations, environmental remediation technologies, photoelectrochemical water-splitting devices, biofuel cells, batteries and supercapacitors.

Dr. Tauseef Ahmad Rangreez is working as a postdoctoral fellow at National Institute of Technology, Srinagar, India. He completed his Ph.D in Applied Chemistry, from Aligarh Muslim University, Aligarh, India on the topic "Development of Nanostructure Organic-Inorganic Composite Materials based Sensors for Inorganic Pollutants". He worked as a Project Fellow under the UGC Funded Research Project entitled "Development of Nanostructured Conductive Organic Inorganic Composite Materials based sensors Functionalities for Organic and Inorganic Pollutants". He completed his Masters in Chemistry from Jamia Hamdard, New Delhi. He has published several research articles of international repute. He has edited books with Springer and Materials Research Forum LLC, U.S.A. His research interest includes ion exchange chromatography, development of nanocomposite sensors for heavy metals and biosensors.

Prof. Abdullah M. Asiri is the Head of the Chemistry Department at King Abdulaziz University since October 2009 and he is the founder and the Director of the Center of Excellence for Advanced Materials Research (CEAMR) since 2010 till date. He is the Professor of Organic Photochemistry. He graduated from King Abdulaziz University (KAU) with B.Sc. in Chemistry in 1990 and a Ph.D. from University of Wales, College of Cardiff, U.K. in 1995. His research interest covers color chemistry, synthesis of novel photochromic and thermochromic systems, synthesis of novel coloring matters and dyeing of textiles, materials chemistry, nanochemistry and nanotechnology, polymers and plastics. Prof. Asiri is the principal supervisors of more than 20 M.Sc. and six Ph.D. theses. He is the main author of ten books of different chemistry disciplines. Prof. Asiri is the Editor-in-Chief of King Abdulaziz University Journal of Science. A major achievement of Prof. Asiri is the research of tribochromic compounds, a new class of compounds which change from slightly or colorless to deep colored when subjected to small pressure or when grind. This discovery was introduced to the scientific community as a new terminology published by International Union of Pure and Applied Chemistry (IUPAC) in 2000. This discovery was awarded a patent from European Patent office and from UK patent. Prof. Asiri involved in many committees at the KAU level and on the national level. He took a major role in the advanced materials committee working for King Abdulaziz City for Science and Technology (KACST) to identify the national plan for science and technology in 2007. Prof. Asiri played a major role in advancing the chemistry education and research in KAU. He has been awarded the best researchers from KAU for the past five years. He also awarded the Young Scientist Award from the Saudi Chemical Society in 2009 and also the first prize for the distinction in science from the Saudi Chemical Society in 2012. He also received a recognition certificate from the American Chemical Society (Gulf region Chapter) for the advancement of chemical science in the Kingdome. He received a Scopus certificate for the most publishing scientist in Saudi Arabia in chemistry in 2008. He is also a member of the editorial board of various journals of international repute. He is the Vice- President of Saudi Chemical Society (Western Province Branch). He holds four USA patents, more than one thousand publications in international journals, several book chapters and edited books.

www.ingramcontent.com/pod-product-compliance
Lightning Source LLC
Chambersburg PA
CBHW071657210326
41597CB00017B/2231